Lecture Notes in Mobility

Series editor

Gereon Meyer, Berlin, Germany

More information about this series at http://www.springer.com/series/11573

Carolin Zachäus · Beate Müller
Gereon Meyer
Editors

Advanced Microsystems for Automotive Applications 2017

Smart Systems Transforming the Automobile

Springer

Editors
Carolin Zachäus
Future Technologies and Europe
VDI/VDE Innovation + Technik GmbH
Berlin
Germany

Gereon Meyer
Future Technologies and Europe
VDI/VDE Innovation + Technik GmbH
Berlin
Germany

Beate Müller
Future Technologies and Europe
VDI/VDE Innovation + Technik GmbH
Berlin
Germany

ISSN 2196-5544 ISSN 2196-5552 (electronic)
Lecture Notes in Mobility
ISBN 978-3-319-66971-7 ISBN 978-3-319-66972-4 (eBook)
DOI 10.1007/978-3-319-66972-4

Library of Congress Control Number: 2017951406

Printed on acid-free paper

This Springer imprint is published by Springer Nature
The registered company is Springer International Publishing AG
The registered company address is: Gewerbestrasse 11, 6330 Cham, Switzerland

Preface

The ambitious goals of climate protection, energy efficiency, air quality, and road safety in combination with current trends in society and economy require restless innovation in sustainable and inclusive road transport solutions. Recent breakthroughs in electric and electronic components and systems, clean propulsion, automation, connectivity, IT-driven business models, and user-centric design promise viable opportunities for the automobile to meet those challenges.

However, it is foreseeable that additional principles will even further transform mobility in the next 5–10 years: Big data analysis, e.g., is already crucial for automation and will further enable cities to better respond to traffic issues, the Internet of Things will improve connected driving and together with blockchain facilitate mobility-as-a-service, robots and artificial intelligence will make human drivers obsolete not only in passenger cars but also potentially in all road vehicles, 3D printing will not only allow the production of novel vehicles but also change the demand for transport by making it possible to replace shipping products by shipping materials and local production, virtual and augmented reality may reduce the need to travel physically, and carbon fiber structures may make the automobile extremely lightweight and efficient. Moreover, maybe in 10–15 years, even more technical revolutions will arise not only creating completely new opportunities from integration and synergetic combinations of these various technologies with electrification, automation and sharing concepts that impact the demand for road transport but also widen the possibilities for road vehicles.

Currently, one of the main topics raising considerable attention of politicians and engineers alike are connected and automated vehicles and their enabling technologies. At the same time, governments are debating on the legal and infrastructural preconditions of automated driving and about how to harmonize the necessary investments. This field faces two major challenges.

First, R&I efforts need to be shifted from proof-of-concept to proof-of-reliability on the system level of automated driving technology. For instance, the performance envelope of sensors, data fusion, and object recognition systems has been pushed considerably in recent years, but does still not cover the complexity that a vehicle encounters in everyday life. Particularly for applications in urban environments and at higher levels of automation, perception of the driving environment is a challenging task to be mastered in a robust fashion. Smart systems integration and connectivity have to play an important role in this domain, with functional safety, fail-operational capabilities and user comfort being of primary concerns.

Second, it is obvious that the economically viable large-scale rollout of driving automation requires agreements on framework conditions between a large and heterogeneous group of stakeholders encompassing the automotive, IT, and telecom sectors as well as road infrastructure providers, public authorities and users.

To this end, two EU-funded Coordination and Support Actions "Safe and Connected Automation of Road Transport (SCOUT)" and "Coordination of Automated Road Transport Deployment for Europe (CARTRE)" are supporting the involvement of different stakeholders and will develop cross-sectorial roadmaps regarding the implementation of high-degree connected and automated driving in Europe. These may serve as blueprints for research and innovation funding and regulatory actions as well as support European Technology Platforms and a wide range of activities regarding connected and automated driving.

The International Forum on Advanced Microsystems for Automotive Applications (AMAA) has been exploring the technological foundations of new paradigms for many years. Consequently, the topic of the twenty-first edition of AMAA, held in Berlin on September 25 and 26, 2017 was "Smart Systems Transforming the Automobile". The AMAA organizers, VDI/VDE Innovation + Technik GmbH together with the European Technology Platform on Smart Systems Integration (EPoSS), greatly acknowledge the support given for this conference.

The papers in this book, a volume of the Lecture Notes in Mobility book series by Springer, were written by leading engineers and researchers who have attended the AMAA 2017 conference to report their recent progress in research and innovation. The paper proposals were peer-reviewed by the members of the AMAA Steering Committee. As the organizers and the chairman of the AMAA 2017, we would like to express our great appreciation to all the authors for their high-quality contributions to the conference and also to this book. We would also like to gratefully acknowledge the tremendous support we have received from our colleagues at VDI/VDE-IT.

Berlin, Germany
July 2017

Carolin Zachäus
Beate Müller
Gereon Meyer

Organisation Committee

Funding Authority

European Commission

Supporting Organisations

European Council for Automotive R&D (EUCAR)
European Association of Automotive Suppliers (CLEPA)
Strategy Board for the Automobile Future (eNOVA)
Advanced Driver Assistance Systems in Europe (ADASE)
Zentralverband Elektrotechnik- und Elektronikindustrie e.V. (ZVEI)
Mikrosystemtechnik Baden-Württemberg e.V.

Organisers

European Technology Platform on Smart Systems Integration (EPoSS)
VDI/VDE Innovation + Technik GmbH

Steering Committee

Mike Babala, TRW Automotive, Livonia MI, USA
Serge Boverie, Continental AG, Toulouse, France
Geoff Callow, Technical & Engineering Consulting, London, UK
Wolfgang Dettmann, Infineon Technologies AG, Neubiberg, Germany
Kay Fürstenberg, Sick AG, Hamburg, Germany
Wolfgang Gessner, VDI/VDE-IT, Berlin, Germany
Roger Grace, Roger Grace Associates, Naples FL, USA

Klaus Gresser, BMW Forschung und Technik GmbH, Munich, Germany
Riccardo Groppo, Ideas & Motion, Cavallermaggiore, Italy
Jochen Langheim, ST Microelectronics, Paris, France
Günter Lugert, Siemens AG, Munich, Germany
Steffen Müller, NXP Semiconductors, Hamburg, Germany
Roland Müller-Fiedler, Robert Bosch GmbH, Stuttgart, Germany
Andy Noble, Ricardo Consulting Engineers Ltd., Shoreham-by-Sea, UK
Pietro Perlo, IFEVS, Sommariva del Bosco, Italy
Christian Rousseau, Renault SA, Guyancourt, France
Jürgen Valldorf, VDI/VDE-IT, Berlin, Germany
David Ward, MIRA Ltd., Nuneaton, UK

Conference Chair

Gereon Meyer, VDI/VDE-IT, Berlin, Germany

Conference Organizing Team

Carolin Zachäus, VDI/VDE-IT, Berlin, Germany
Beate Müller, VDI/VDE-IT, Berlin, Germany

Contents

Part V Legal Framework and Impact

Part I
Smart Sensors

Smart Sensor Technology as the Foundation of the IoT: Optical Microsystems Enable Interactive Laser Projection

Stefan Finkbeiner

Abstract Consumer electronics such as smartphones, tablets, and wearables are part of our everyday life—visible everywhere and taken for granted. Less visible however are the small MEMS (micro-electromechanical systems) sensors that are an integral part of these devices. Smart sensor technology enables things to be sensed and connected—in all parts of our daily life, in homes, vehicles, cities. With the emergence of the Internet of Things (IoT), more and more devices become connected which results in demanding challenges for MEMS sensor technology providers—in addition to the trends of low cost, small size, low power consumption as well as overall system performance. The exciting developments in the IoT are advancing at an amazing pace. It is not just about how devices communicate or sense their surrounding environments, but increasingly about how technology interacts with human beings. Laser-projected virtual interfaces based on optical MEMS are a new fascinating solution in a world of previously unimaginable opportunities. They give any kind of device a unique personality of its own, enabling technology to interact with people, to make life simpler and more exciting. It is a ground-breaking solution for embedded projectors and augmented reality applications such as games, infotainment as well as in-car head-up displays or intelligent head lamps for automated driving.

Keywords MEMS · Sensor · Internet of things · IoT · Projector · Interactive · Head-up displays · HUD · Wearables

S. Finkbeiner (✉)
Bosch Sensortec GmbH, Gerhard-Kindler-Straße 9, 72770 Reutlingen/Kusterdingen, Germany
e-mail: stefan.finkbeiner@bosch-sensortec.com

C. Zachäus et al. (eds.), *Advanced Microsystems for Automotive Applications 2017*, Lecture Notes in Mobility, DOI 10.1007/978-3-319-66972-4_1

1 MEMS Sensors—The Hidden Champions

MEMS (micro electromechanical systems) sensors are a key technology for the mobile and connected world. These compact electronic sensors are the hidden champions of everyday life, providing rich data used in a huge variety of applications, such as motion tracking, temperature sensing, and many others.

1.1 Enablers for the Internet of Things

Smart sensor technology enables things to be sensed and connected—in all parts of our daily life, in homes, vehicles, cities. With the emergence of the Internet of Things (IoT), more and more devices become connected which results in demanding challenges for MEMS sensor technology providers—in addition to the trends of low cost, small size, low power consumption as well as overall system performance.

To give just a few examples of where MEMS sensors are finding uses with the IoT:

- Parking spot detection
- Indoor and outdoor navigation
- Indoor air quality
- Sleep monitoring
- Intrusion detection
- Asset tracking
- Augmented reality
- Step counting
- Calorie tracking

In the IoT, everything will be connected. Today about 6 billion devices are connected worldwide, according market experts (source: Gartner). By 2020, this figure will have grown to 21 billion, with the global market for IoT solutions expected to be worth US $250 bn.

1.2 Challenges and Barriers for IoT Sensors

The IoT is technologically demanding for sensors. While there are many technologies available, they are not always adapted for IoT applications. In particular, power consumption must be low for always-on applications, and sensors must meet tough demands on size, scalability and cost.

There is also need for a high degree of customization, across a huge range of different applications in the home, vehicle, and city, and across many different industries. There is a lack of synergies and standardization across applications.

This means that vendors need deep application know-how, and the ability to meet the needs of low-volume customers.

The IoT is complex, which means cooperation and collaboration is needed between vendors. The value is in end-to-end solutions, so by establishing an ecosystem, companies can work together to cut time to market and deliver the right solutions.

1.3 The Role of Smart Sensors in the IoT

Smart sensor technology is the foundation of the IoT.

Highly integrated smart sensor hubs can handle the requirements of multiple, complex environments, and simplify the application design process.

Smart sensor hubs can overcome three key challenges:

- Technology: leveraging core MEMS and system know-how, to provide embedded intelligence in compact, high-performing devices, with low power consumption.
- Fragmentation: leveraging application knowledge, to provide an integrated solution including hardware and software, and in particular application-specific software
- Complexity: providing simple, turnkey solutions, and co-operating with third parties to deliver reference designs

For example, Bosch Sensortec's integrated sensor hubs BHI160 and BHA250 combine our lowest power sensors, best-in-class sensor data fusion software and an optimized microcontroller, to provide the lowest power solution without compromising features or performance.

2 Interactive Laser Projection

The IoT user interface is playing an increasingly important role in determining the type and amount of user input and the type and form of feedback received from technology.

We are essentially viewing the world through one screen or another. With our heads tilted down, we often live our lives inside of our smart devices, where smartphones eat up our time and, in a way, present an irresistibly addictive immersive environment for our information hungry senses.

Of course, there are many situations where a smartphone provides a very powerful user interface (UI) for accessing the active processes within our small part of the IoT universe. Nevertheless, we foresee many scenarios where the user interface will become indiscernible from the real and tangible world around us, appearing on demand within the context of where it is intuitively expected, but remaining hidden from sight when not required.

What a refreshing innovation it would be if our devices could project information straight out of the digital space on to the real world, thereby expanding the scope of our interaction with the IoT domain! This is now termed as "Mixed Reality".

The world is awash with sensors, devices and IoT applications, but it is our contention that the way that we interact with technology is currently undergoing a silent revolution.

In the upcoming decade, we foresee interactive laser projection fundamentally transforming our present concept of the user interface, of the way we technologically interact with our world. The exciting news is that interactive projection is here and now, no longer just on the movie screen—with the BML050 laser projection microscanner from Bosch Sensortec it has now become reality (Fig. 1).

2.1 Making User Interfaces Simpler, More Flexible … and More Fun

Let us take for example, a standard consumer appliance like an espresso machine. A high-end machine may have so many functions and features that the user manual

Fig. 1 Interactive projection with microscanner

is more complicated to navigate than Shanghai in rush hour—buttons, pop-up or drop-down menus, sliders, dials, knobs, lights, buzzers, radio tuner, etc., are we not just making coffee? And so, overwhelmed by complexity, the user leaves this modern monster gathering dust on the countertop and makes instant coffee!

With interactive projection, physical input/output elements are practically eliminated, thus fundamentally simplifying your espresso machine, toaster, or dishwasher. Making appliances and devices smaller and more intuitive, eliminating clutter and distractions will dramatically raise their appeal to the consumer. The projected user interface dynamically adapts and responds to specific user tasks, showing only relevant task-specific information. This means that complex, variable content is presented intuitively in a clean and easy-to-understand format. The coffee machine can now project a large, legible image on to any surface like a kitchen countertop or table.

The user needs only tap the projected button or menu item to set their preferred coffee specs. What is more, the user's interaction with technology is accompanied by the intuitive assistance of the system itself, making daily tasks more fun and hassle-free.

The display now functions even when the surface is obstructed or dirty, for instance when cooking. No more greasy fingerprints on your fine polished appliances—there is no need to physically touch them anymore, they will stay clean. In addition, the user can watch recipe videos on a virtual screen or chat with friends using a virtual keyboard.

2.2 *Interactive Projection in Practice*

To make all of this a reality, Bosch Sensortec has developed the BML050, a MEMS microscanner housing two MEMS mirrors, which when integrated into a complete projector reference design, project an RGB color laser image on to any kind of surface (Fig. 2).

Beyond laser projection, the BML050 also features the essential gesture sensing function. By measuring the optical feedback from illuminated objects, the system detects any interference between the user and the projected content. Since the optical feedback is determined by the position of the scanner, no calibration between projector and detection is needed. Just switch it on and press a button.

2.3 *A Window to the IoT*

Mixed Reality based on interactive projection will be utilized in all manner of devices, limited only by the design engineer's imagination.

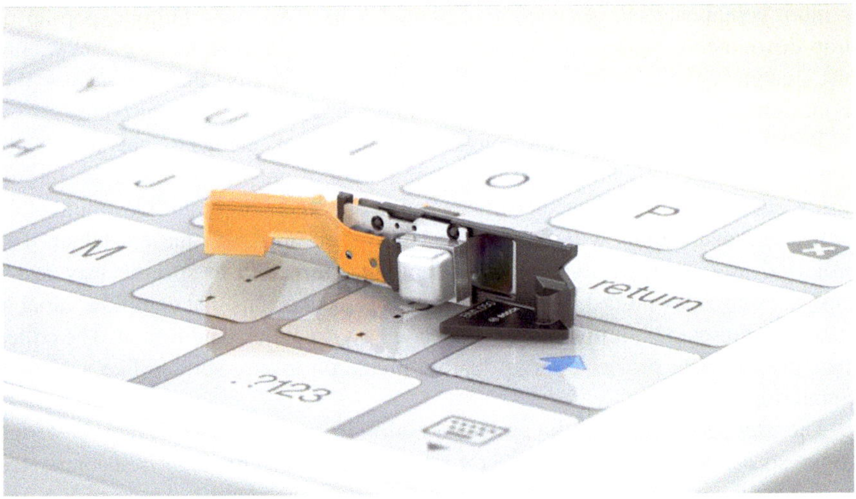

Fig. 2 BML050 microscanner reference design

Furthermore, a complex UI environment consisting of dozens of individual screens and buttons poses a definite barrier to broader IoT adoption. Consumers simply cannot handle many more screens and buttons installed around their homes. Interactive projection provides a way to create on-demand interfaces, providing larger, clearer, more intuitive control elements that are hidden from sight when not required. The envisioned result is that homes will not be cluttered with any more screens and control panels—a clean wall approach—a connected world without any light switches, buttons or LCD displays.

Devices such as cooking appliances or service robots will be characterized by simple, clean design concepts. Usability will be much improved and manufacturers will be free to add many more value-adding features to their products without being limited by the physical interface.

For consumers, a context-related, menu guided UI will bring significant benefits. They can throw out their printed user manuals as intuitive help will always be on hand to guide them through the user interface, including video and images where required. Providing visual explanations makes devices substantially easier to use, particularly for people with an aversion to reading user manuals but in particular for elderly people or children facing the problem of how to get started with their new service robots. Interactive projection will complement the voice-controlled user interfaces that are currently penetrating the consumer electronics market.

Moreover, the projected image will not be limited only to the user interface function. It will be used to play videos and project graphics and texts, for example: recipes and YouTube cooking videos. It will also provide a way to access social media or make video calls to neighbors and friends.

2.4 *Interactive Projection for the Automotive Industry*

In the automotive industry, there are also many potential applications for interactive projection, with low cost integrated modules.

Head-up displays (HUD) and driver information can benefit from interaction, and other devices in the car can use projection for the human–machine interface, both for the driver and for passengers. Projected display can be accommodated also on curved surfaces, so that new design freedoms are opened up.

While HUDs today largely are reserved to the premium segment, it will be possible in the future to offer such systems even in compact and subcompact models with limited space and cost constraints, thus making an important contribution to improving road safety.

Interactive projection will also find uses in the more technical side of the industry, such as in car manufacture and repair.

2.4.1 Industry Teamwork

Leading manufacturers of light sources and micro-mechanical components, as well as a carmaker and an automotive supplier, have teamed up with a research organization to create the basics for highly integrated, economically producible laser projection modules for use in volume markets.

The research project PICOLO uses the results obtained within the framework of the Federation of MOLAS, in particular relating to green and blue laser diodes, as well as to the use of a MEMS scanner for the laser projection.

Very low cost is critical to achieve high market penetration, as well as highly integrated full-color-laser-projectors with power-saving driver electronics that can be manufactured in volume-capable production processes.

To achieve this goal, the development of high-efficient direct-green laser diodes is within the joint research project PICOLO, as well as developing suitable MEMS scanner and driver electronics, specific laser drivers, and high volume production capable, innovative concepts for integration.

Building upon initial works, Robert Bosch GmbH and BMW Forschung und Technik GmbH were involved. These partners will investigate the use of laser projection in low cost, building space reduced HUD modules, to the auto-stereoscopic 3D projection for the contact-analog display of information for driver assistance systems.

2.5 *Wearables and Beyond*

Interactive projection will also find many niches in wearables and other portable devices. It will play a key role in resolving the present dilemma of users wanting tiny, lightweight devices, while requiring large screens for managing increasingly sophisticated apps. The BML050 microscanner is compact enough for installation in portable smart devices.

The microscanner will most probably be heavily utilized in gaming to integrate mixed reality with projected images on real objects. Of course, navigation applications are a natural match, where the phrase of 'knowing an area like the back of your hand' will take on a new, much more literal meaning.

As this technology develops further, new applications that are still beyond what we can imagine today will emerge and soon become a natural part of our everyday lives. Perhaps 3 years from now, we might see smart glasses with a tiny interactive MEMS projector integrated in the frame. The wearer will see an image projected on to the lens providing information or augmented reality.

2.6 A Compact Module

While the concepts described above are still on the drawing board, the MEMS mirror technology enabling them is not. In 2017, the BML050 has been provided to selected customers, and will most likely go mainstream in 2018.

The BML050 microscanner is small and very power efficient, making it suitable for mobile and battery-powered devices. It delivers outstanding projection quality with advanced speckle reduction and precise control of the MEMS mirrors and laser

Fig. 3 BML050 microscanner reference design

diodes. Projection is focus-free, and its vivid laser colors exceed the natural color spectrum perceptible to the human eye.

A reference design will be available for evaluation by the second half of 2017. This microscanner is a fully aligned, ready-to-go calibrated solution, which includes all the MEMS mirrors and control circuits, lasers and laser drivers, and power management in a single compact package (Fig. 3).

3 Conclusion

With the Bosch Sensortec BML050 MEMS scanner, engineers will transform how we create brand new user interfaces in many products. Interactive projection will make daily life much easier and more exciting for consumers and will remove many of the present barriers to IoT adoption, including many automotive applications.

At this stage, the technology is ready for mainstream implementation, and we are looking forward to seeing where designers will take it. We foresee an exciting period of innovation ahead.

Unit for Investigation of the Working Environment for Electronics in Harsh Environments, ESU

Hans Grönqvist, Per-Erik Tegehall, Oscar Lidström, Heike Wünscher, Arndt Steinke, Hans Richert and Peter Lagerkvist

Abstract When electronic equipment is used in harsh environments with long expected lifetime, there is a need to understand that environment more in detail. This situation is today a reality for many application areas including the automotive sector, heavy industry, the defense sector, and more. To fully understand the working environment, a unit has been developed to monitor physical data such as temperature, vibration, humidity, condensation, etc., to be used in the product development phase for new products. This paper presents the underlying principles for the ESU (Environmental Supervision Unit) and details on the design.

Keywords Harsh environment · Monitoring unit · HALT · Condensation sensor · Multi sensor unit · Reliability

H. Grönqvist (✉) · P.-E. Tegehall · O. Lidström
Swerea IVF, Box 104, 431 22 Mölndal, Sweden
e-mail: hans.gronqvist@swerea.se

P.-E. Tegehall
e-mail: per-erik.tegehall@swerea.se

O. Lidström
e-mail: oscar.lidstrom@swerea.se

H. Wünscher · A. Steinke
CiS Forschungsinstitut für Mikrosensorik GmbH, Konrad-Zuse-Str. 14, 99099 Erfurt, Germany
e-mail: hwuenscher@cismst.de

A. Steinke
e-mail: asteinke@cismst.de

H. Richert
SETEK Elektronik AB, Krokslätts Fabriker 26, 431 37 Mölndal, Sweden
e-mail: h.richert@setek.se

P. Lagerkvist
Niranova AB, Krokslätts Fabriker 26, 431 37 Mölndal, Sweden
e-mail: peter.lagerkvist@niranova.se

C. Zachäus et al. (eds.), *Advanced Microsystems for Automotive Applications 2017*, Lecture Notes in Mobility, DOI 10.1007/978-3-319-66972-4_2

1 Introduction

The basic idea behind the project is to provide the industry with a unit for monitoring the environment for the customers' electronics units during the development phase and/or during actual use. This ESU (Environmental Supervision Unit) can be configured in many different ways using transducers for different physical parameters. The rationale for this is that in order to design electronics equipment that needs to be reliable in harsh environment, there is a need to understand this environment in detail.

Reliability of an electronic product is normally assured by designing, manufacturing, and testing them according to standards. These standards are based on best practice of mature technologies and the main focus is to assure good manufacturing quality. However, new types of components such as QFN and fine-pitch BGA components may have inadequate solder joint life even if they have been produced with perfect manufacturing quality. This is mainly a problem for high reliability products that are expected to have a long life in harsh conditions. The main factors affecting the life of solder joints are thermomechanical stress due to temperature variations and mechanical stress due to vibration or shock. Therefore, in order to assess the life of solder joints it is necessary to have information of the stresses solder joints will be exposed to during the whole life of the product, not only during service but also during handling and storage. It is not only the maximum and minimum temperatures and the maximum vibration levels that need to be determined. The number of temperature cycles with different delta temperature and the variation in vibration levels must be known in order to facilitate assessment of the life of the solder joints.

In addition, condensation of water on the board of electronic equipments is a well-known cause of failures.

In an ongoing Swedish national project, "Requirements specification and verification of environmental protection and life of solder joints to components", the ESU will be used for characterizing the actual field environment for a number of products used in harsh environments. This will give information of the usefulness of the ESU. The project is supported by the Swedish Governmental Agency for Innovation Systems (VINNOVA).

2 Monitoring Unit, ESU

The SME Setek Elektronik AB located in Gothenburg, Sweden, is currently investigating the market potential for the ESU and will build a number of prototypes to be evaluated by an industrial reference group. This group consists of vehicle manufacturers, (cars, trucks, trains, boats, etc.) in the project "Requirements specification and verification of environmental protection and life of solder joints to components" mentioned earlier.

The transducers for the current version of the ESU is equipped with

- Condensation sensor provided by CiS. (A unique sensor on the market)
- Several sensors for temperature. (Commercial)
- Sensor for relative humidity. (Commercial)
- Sensor for vibration, acceleration, and displacement. (Commercial)
- A large memory for storing environmental data for long times

Figure 1 shows the current design and Fig. 2 shows that particular sensor. A more detailed layout of the sensor chip is shown in Fig. 3.

The basic functionality of the condensation sensor is: Condensation leads to the formation of droplet of the sensor surface. The detection of droplet bases on the capacitive principle using an electric stray field on a dense surface. The principle is shown in Fig. 4. The packaging withstands the corrosive properties of condensate over many years. The upper passivation layer consists of silicon nitride, is mechanical stable, and allows a cleaning if necessary. It enables the detection of droplets as small as some μm on the sensor surface. The overall impedance of the stray field capacity shows a marked frequency response, which is greatly dependent

Fig. 1 Picture of the prototype

Fig. 2 Picture of the unique condensation sensor

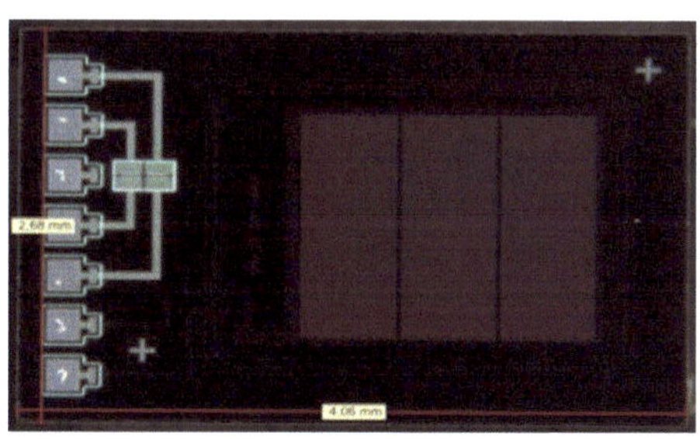

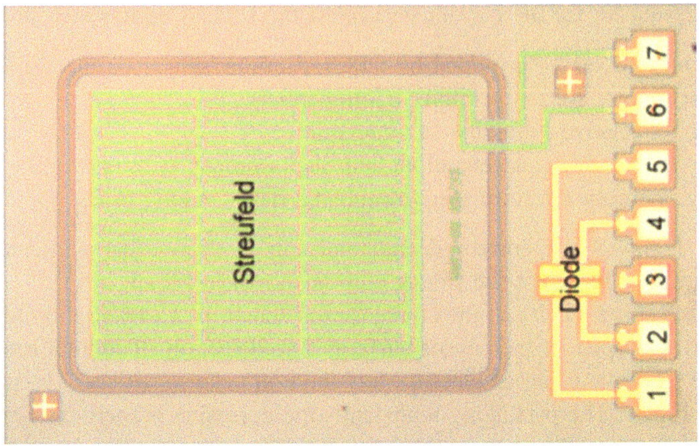

Fig. 3 Layout of the condensation sensor chip

Fig. 4 Measurement principle of the condensation sensor

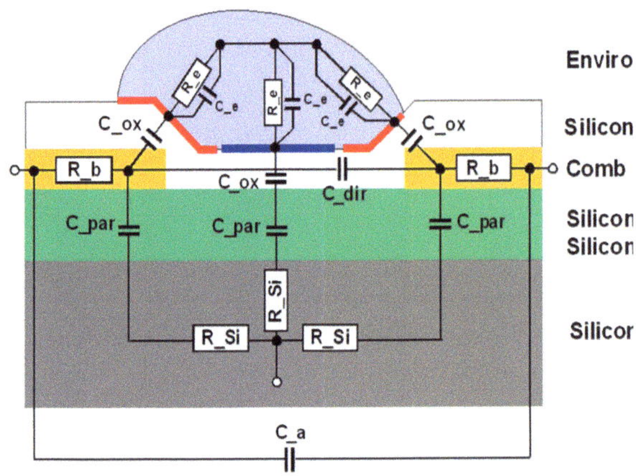

in its course on the properties of the surrounding medium. Figure 5 shows a typical packaged condensation sensor.

If necessary, the extent of the condensation can be given quantitatively (Fig. 6). This requires that the sensor is calibrated.

A capacity frequency converter allows the transportation of the signal via a longer distance. In principle, the method is also appropriate to give information on impurities [1]. The sensor also includes a diode for measuring the temperature which allows detecting the condensation and temperature at the same place. Figure 7 shows the characteristics of seven different temperature diodes of one type.

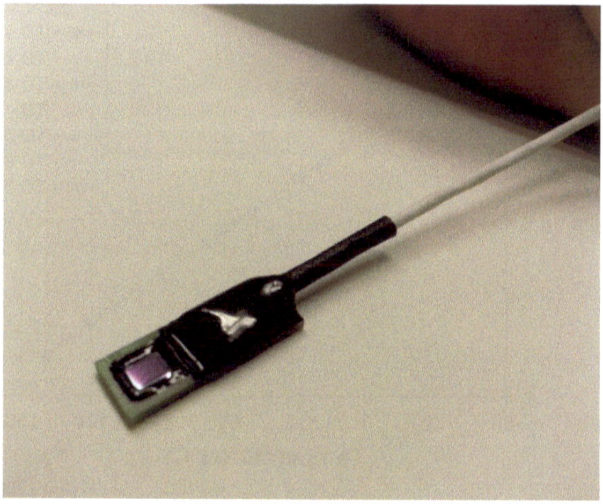

Fig. 5 Packaging of the condensation sensor

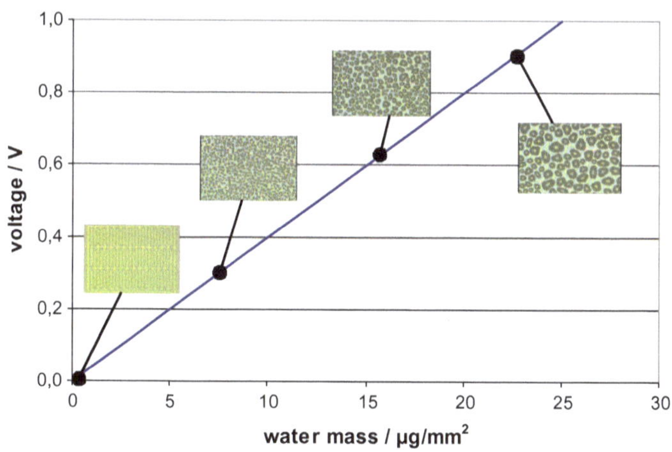

Fig. 6 Correlation of droplet size and signal

The condensation sensor is mounted on a tinned pad on the ESU with a thermally conductive epoxy (Fig. 8). The pad is connected to the underside of the ESU FR4-card by thermal vias to increase the heat flow from the mounting surface to the sensor.

The ESU can be mounted to any surface either by a screw through the hole in the middle of the device or by a differential tape. The adhesion is strongest on the ESU side of the tape which facilitates removal or relocation of the ESU by not leaving any residues on the mounting surface. The tape can easily be replaced by a new one upon removal of the ESU.

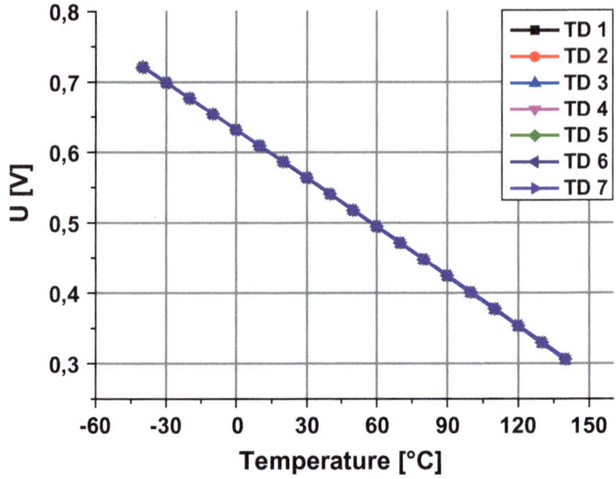

Fig. 7 Voltage–temperature characteristics of seven different temperature diodes of one type

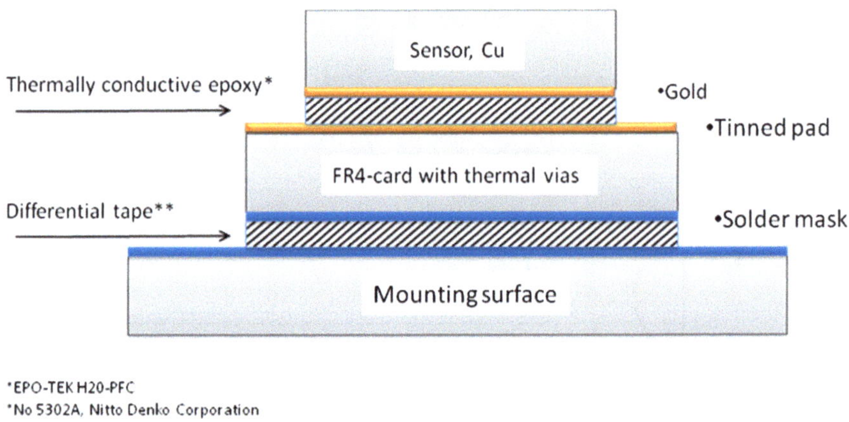

Fig. 8 Mounting of the condensation sensor and the ESU in an application

Since the ESU will be monitoring harsh environments the prototype will need to function in that environment.

The design specifications in short for the unit are

- Temperature range: -40 – +125 °C
- Temperature ramp: 60 °K/min
- Current consumption: as low as practical, few mA at 12 V.
- Supply voltage range: 7–36 V DC
- Vibration level: maximum 65 G RMS.

It should be observed that some of these requirements are operational and some are based on survival of the unit. We can, for instance, not measure vibration levels up to 65 G RMS with the current accelerometer but the sensor survive that level without damage.

2.1 *ESU Main Data*

The main data for the ESU in the current configuration are

- Diameter: 46 mm.
- Height: 12 mm
- Weight: 10 g
- Processor ARM Cortex M0+ from NXP.
- 256 M bit NOR flash
- Minimum 100,000 ERASE cycles per sector
- More than 20 years data retention
- Current consumption while measuring: 12–13 mA typical @13.6 V

It should be emphasized that the ESU can be configured in a modular way with additional sensor types that a customer should require. Also, times between measurement samples, recording times for each measurement, etc., are relatively simple to adjust in the software of the unit.

2.1.1 Condensation Measurement

The condensation sensor measures micro-condensation and presents the result as volume precipitated water per surface unit.

One new sample is taken every 100 ms.

To have a reliable analogue value requires manual calibration of each individual sensor. The measurement in the ESU is therefore "digital" and will just detect condensation or no condensation. The sensor will trig when condensation drops are approximately 3–4 m or larger.

The sensor element should as far as possible be protected from contamination.

It is very important that the sensor has the same temperature as the test surface.

The condensation sensor is normally glued to the ESU but as an option it can be delivered mounted on a cable at a distance of up to 90 cm (or longer). If this option is selected, one of the external temperature sensors must be placed close to the condensation sensor.

2.1.2 Relative Humidity Measurement

Relative humidity is measured using a commercially available sensor with the following data:

- Relative Humidity Accuracy ± 2% (typical)
- Integrated temperature accuracy ± 0.2 °C (typical)
- 14 Bit Measurement Resolution
- One new sample taken every 100 ms.

2.1.3 Vibration Measurement

A commercially available accelerometer is used for measuring vibrations.

- 3 axis measurement
- ±16 G range
- 13-bit resolution
- Max data output rate = 3200 Hz - > Max bandwidth = 1600 Hz. Not fully utilized at this time.

One measurement value for each axis is produced for every sample.

It is important that the ESU is firmly connected to the test object to get reliable acceleration values especially if frequencies are high.

One new sample taken every 1 ms. The maximum value measured during the selected sample time is stored during log. A more sophisticated calculation could be used in future.

2.1.4 Temperature Measurement

Up to five NTC temperature sensors can be connected to the ESU and placed at different spots on the test object. There is also a sensor placed close to the condensation sensor on the ESU.

- Temperature range: −55 to 155 °C
- Inaccuracy: TBD

New samples are taken every 100 ms.

2.1.5 RTC

A real-time clock is integrated to the ESU to be able to time tag each measurement record.

A super capacitor is used to supply the RTC when normal supply is removed.

- Clock inaccuracy: ± 10 ppm (typical)
- Time tick: 1 s
- Max run time without supply: TBD

The real-time clock is important since often there is a need to back track when a particular event happened, like when condensation occurred. The real-time clock provides a mean for synchronizing data from the ESU to other measurements done in the evaluation of a product.

2.1.6 User Interface

A standard asynchronous serial connection (UART) is used to give commands to the ESU and to read out data. Any terminal program with logging capability could be used.

Data is sent out on the serial port in a CSV (Comma Separated Values) format which can easily be loaded into Excel and other programs.

2.2 *Reliability of the ESU*

The unit has been tested using HALT (highly accelerated life testing) which is a stress test used to quickly find defects and weaknesses related to the product design, manufacturing processes, and material selection of the product. During the HALT, stresses are applied to the unit by step-wise increasing levels of vibration and/or temperature outside the p roduct specification in order to cause failures in short time. When failures are found, proper actions can be taken to increase the products reliability.

The halt consists of four parts; *Thermal Step Stress Test*, During which the ESU was tested between −90 and 125 °C, *Rapid Thermal Transition Test* where the unit is exposed to rapid thermal chocks between −40 and 125 °C with a ramp of at least 60 °C/min, *Vibration Step Stress Test* where the ESU was subjected to omni-axial random vibrations up to 65 G_{rms} and a *Combined Environments Test* where the *Rapid Thermal Chock* test is combined with vibrations up to 65 G_{rms}.

Findings from the HALT showed that the ESU had a robust design but with potential for improvements of the reliability by stabilizing its Super Capacitor mechanically. Later editions of the ESU have been ruggedized by gluing the Super Capacitor to the PCB.

2.3 *EMC Test*

The emission of the ESU was tested using standard test equipment in the EMC-lab of Swerea IFV. Since there is a broad application area, the unit was tested applying the limits for household applications from CISPR 32 which are the tougher requirements. The unit passed these tests.

Depending on the actual working environment of the unit, there could be requirements to test the unit also according to CISPR 25 with the adoptions to that standard following the different companies' interpretations. This would require a different setup of the test equipment. It is foreseeable that these tests will be crucial for different fields of application.

Immunity to electromagnetic fields and pulses has not been tested at this stage. Here, the required tests will also be determined by the actual working environment according to the end users' needs.

3 Market Assessments

The first batch of prototypes was delivered to an industrial reference group for evaluation in the spring 2017. This group consists mainly of companies in the automotive sector in Sweden. One initial result was that the configuration of the unit had to be adapted to suit each customer regarding what sensors where of interest and details in the software and how the results should be gathered. First results from the field tests are expected early autumn.

The ESU was also presented on a trade show in Gothenburg in March 2017 and gained some interest among companies in business sectors not initially foreseen. These applications were in ventilation systems for buildings and general environmental monitoring in housing regarding for instance condensation. For the SME, Setek Elektronik AB these applications were the first commercial successes for the unit.

The ESU will also be used in a national research project regarding monitoring of the environment for fresh water plants in Sweden.

Apart from the industrial reference group, the SME also works on marketing of the ESU towards the defense industry. In Sweden the organization FMV (Swedish Defence Material Administration) is of interest.

Acknowledgements This project received funding from:

The Innovation Action Smarter-SI receives funding from the European Community's Programme HORIZON 2020 under GA No. 644596 and from the Swiss State Secretariat for Education, Research and Innovation (SERI) under contract number 15.0085. SMARTER-SI is part of the Smart Anything Everywhere Initiative of the European Commission.

Reference

Westenthanner M, Barthel A, He P, Beckmann D, Steinke A, Tobehn I, Frank T., Pliquett U (2013) Assessment of suspension medium conductivity by means of micro electrodes. J Phys Conf Series 434. article id. 012093

Automotive Synthetic Aperture Radar System Based on 24 GHz Series Sensors

Fabian Harrer, Florian Pfeiffer, Andreas Löffler, Thomas Gisder and Erwin Biebl

Abstract This paper presents a Synthetic Aperture Radar (SAR) system for automotive applications. The focus is on the use of current series-sensor technology. Typical sensors' parameters are discussed and their effect on SAR image quality is displayed. A complete model for automotive SAR calculation is also presented. Critical properties of SAR in combination with automotive setups are illustrated. Measurements and simulations for a parking-lot scenario created with that model are shown. The system is able to distinguish between cars and parking slots.

Keywords SAR · Radar · ISM band · Azimuth resolution

F. Harrer (✉) · F. Pfeiffer
Perisens GmbH, Lichtenbergstraße 8, 85748 Garching Bei München, Germany
e-mail: harrer@perisens.de

F. Pfeiffer
e-mail: pfeiffer@perisens.de

A. Löffler
ADC Automotive Distance Control Systems GmbH, Peter-Dornier-Straße 10, 88131 Lindau, Germany
e-mail: Andreas.3.Loeffler@continental-corporation.com

T. Gisder
Volkswagen Aktiengesellschaft, Brieffach 1777, 38436 Wolfsburg, Germany
e-mail: thomas.gisder@volkswagen.de

E. Biebl
Technical University of Munich, Arcisstraße 21, 80333 Munich, Germany
e-mail: biebl@tum.de

C. Zachäus et al. (eds.), *Advanced Microsystems for Automotive Applications 2017*,
Lecture Notes in Mobility, DOI 10.1007/978-3-319-66972-4_3

1 Introduction

Driver-assistance systems need a good environment recognition before they begin to actively intervene. Modern upper class cars are equipped with different sensor technologies. They provide information about road surroundings such as range, velocity, and arrival direction. Technologies currently available are radar, optical (camera and lidar), and ultrasonic sensors. Each of these technologies has strengths and weaknesses limiting its field of application. One of radar sensors' main benefits relative to optical systems is the former's robustness and independence from weather and environmental conditions. However traditional radar sensors' small aperture resulting in their low resolution of the direction of arrival or respectively azimuth resolution, is a limitation. As shown in Sect. 4.1, Real Aperture Radar (RAR) sensors can theoretically achieve resolutions of several degrees. This suffices for maneuvers demanding less accuracy, such as lane change or blind-spot detection in dynamic scenarios, because several safety coefficients have to be implemented. These safeties can be reduced for maneuvers such as parking in static scenarios, assuming that measurements with greater accuracy are available.

Synthetic Aperture Radar (SAR) processing is evaluated for automotive scenarios to counter this restriction. SAR is a post-processing algorithm for side-looking radar originally designed for remote sensing applications. It evaluates the radar signal's phase change over the traveled path. This process allows very long synthetic apertures to be created over the vehicle trajectory, which cannot be produced as a real aperture. This has the advantage that azimuth resolution increases inversely with aperture size.

Short-range radars are currently available on the market for side-looking applications such as lane change assist. The aim of the proposed system is to use available sensors with extended capabilities so that no additional hardware is necessary. A high-resolution environment map can be created with it to identify obstacles such as vehicles blocking parking lots or narrow passages. Figure 1

Fig. 1 Mounting position of side-looking radar sensors

shows the mounting position of side-looking radar sensors. The drawback of SAR processing is that it is limited to static scenarios and side-looking applications.

1.1 Automotive Radar Sensors

Two frequency bands are available worldwide for automotive radar: the 24.05–24.25 GHz frequency range (ISM band) used for short- and mid-range sensors and the 76–77 GHz band (ITU-R M1452) used for mid- and far-range sensors. The band from 77 to 81 GHz is also allocated in Europe for short- and mid-range; similar efforts are made in other parts of the world. To estimate the performance of an automotive SAR system, typical parameters are listed in (Winner et al. 2009). Section 4 provides a performance evaluation for all frequency bands and a comparison of SAR to traditional radar.

1.2 Odometry

A precise position measurement is necessary along with coherent radar measurements for SAR calculation since the synthetic aperture is developed over the sensor's trajectory. A nearly linear flight trajectory can be assumed for traditional airborne SAR systems. However this assumption does not hold in automotive systems. Steering maneuvers and velocity variations must be considered in the latter. For testing purposes, highly precise Inertial Measurement Units (IMU) such as the OxTS RT3000 are available. For series systems on the other hand, these devices are economically infeasible. Vehicle odometry based on rotary encoders and gyroscopes can hence be used. The remainder will also illuminate the consequences of inadequate trajectory estimation.

2 Related Work

Several near-range SAR approaches are described in the literature.

(Gerbl 2007) describes a general approach for near-range SAR. The author mentions the scenarios of roadside measurements and the detection of wild animals during pasture mowing.

In (Wu and Zwick 2009), parking-lot scenarios are evaluated on a simulated basis. In (Wu et al. 2011), the same authors subsequently present a motion-compensation algorithm together with parking-lot measurements.

The authors of (Iqbal et al. 2015) also performed measurements in a parking-lot scenario. They use a linear stage with a customized radar with a 5 GHz bandwidth. The range Doppler and line processing algorithm are compared in the evaluation.

A SAR for the near-range is implemented in (Reichthalhammer 2012). The author uses a customized radar in a car with a rotary encoder for motion measurement.

The authors of (Mure-Dubois 2011) use a 77 GHz automotive SAR sensor on a linear rail and compare several SAR algorithms.

Previous papers focus on the creation of SAR images by using laboratory equipment. In this paper, SAR images are produced using series sensors and, if possible, data that is already available in the car. The proposed approach is distinguished from previous ones in that the system is completely integrated into a vehicle. Series sensors in terms of radar and position measurement are used. No compromises have to be accepted concerning weather and temperature conditions, as the sensors are designed for automotive conditions.

3 SAR Algorithm

For most SAR algorithms, such as range Doppler, azimuth operations are performed in Fourier domain, which allows efficient calculation (Cumming and Wong 2005). For this transformation, an equally spaced, linear discrete trajectory is necessary. The Fourier transform has to be calculated for the entire azimuth distance, which precludes a floating (updating) calculation of the image. For the present system, some assumptions from the previous work cannot hold:

- In series sensors, the Pulse Repetition Interval (PRI) cannot be adapted to compensate for velocity changes, so constant distance between azimuth sampling points is unavailable.
- Steering maneuvers must be allowed in urban driving scenarios, so a near-linear trajectory is unavailable.
- SAR images must be available immediately and continuously for driver-assistance actions.

No SAR calculation is possible in the Fourier domain and a time-domain backprojection algorithm is used (Reichthalhammer 2012), which allows the iterative complex addition of phase-corrected samples. The new sample's positions do not need to be linear, so arbitrary steering and non-uniform sampling are possible as long as the spatial sampling theorem (Gerbl 2007) is fulfilled. The image can be independently updated for any new radar measurement. One update of the backprojection algorithm is calculated as follows (Gorham and Moore 2010):

- Calculate the differential range for each pixel in the image
- Calculate phase correction for each pixel.
- Interpolate input data for each pixel's distance.
- Apply phase correction.
- Update on previous step's SAR image.

4 Performance Estimation

The performance expected for SAR systems with typical modern automotive radar sensors will be illustrated below. The resolution limits theoretically possible in the azimuth and range directions are given and compared for the SAR and RAR case. An estimation of possible vehicle velocity is then derived.

4.1 Azimuth Resolution

According to (Skolnik 2008), the azimuth resolution of a Real Aperture (RA) is given by

$$\Delta_{A,RA} = \frac{50.8^\circ \cdot \lambda}{d_{\mathrm{a}}}, \quad (1)$$

where d_a is the antenna's real aperture.

Traditional SAR literature focuses on satellite and airborne SAR systems, which have relatively narrow antenna beam widths. This allows several approximations calculating of the maximum azimuth resolution. In (Klausing et al. 1999), the resolution in the direction of movement of a Synthetic Aperture (SA) is given as

$$\Delta_{\mathrm{A,SA}} = \frac{d_{\mathrm{a}}}{2}. \quad (2)$$

These simplifications do not hold for near-range systems like those in the automotive case with half-power beam widths around 80° for a single channel. According to (Gerbl et al. 2007), the maximum resolution in the azimuth direction depends only on the antenna's beam width γ_{az},

$$\Delta_{\mathrm{A,SA}} = 0.3 \cdot \frac{\lambda}{\sin(\gamma_{\mathrm{az}}/2)}. \quad (3)$$

Figure 2 shows the azimuth resolution over antenna beam width with a marker for the parameters of the sensor under consideration. This resolution is independent of range, in contrast to a RAR, where trapezoidal resolution cells with angular resolution result.

For a typical dimension of automotive radar sensors, this leads to the maximum azimuth resolution given in Table 1. Since the azimuth resolution is independent of range, the angle is referenced to a range of 2 m.

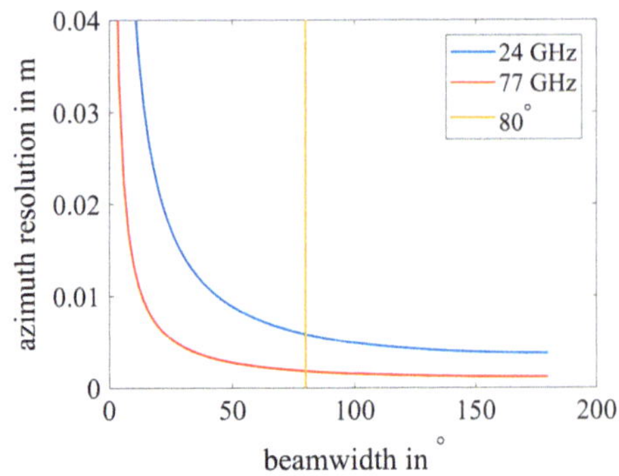

Fig. 2 Azimuth resolution over beam width for SAR

Table 1 Typical automotive radar resolutions

Radar operating frequency band (GHz)	Maximum azimuth resolution	
	RA	SA (2 m)
24.05–24.25	9.0°	5.8 mm ≙ 0.17°
76–77	2.8°	1.8 mm ≙ 0.05°
77–81	2.8°	1.8 mm ≙ 0.05°
Radar operating frequency band (GHz)	Maximum range resolution (cm)	
24.05–24.25	75	
76–77	15	
77–81	5	

4.2 *Range Resolution*

SAR processing does not influence range resolution, as it depends only on the bandwidth. In (Skolnik 2008), the range resolution is given as follows

$$\Delta_R = \frac{c}{2B}. \tag{4}$$

Table 1 shows the range resolution for all frequency bands. It also shows a dramatic mismatch between azimuth and range resolution, especially for the low bandwidth in the 24 GHz band, which results in an hour-glass-shaped point spread function as shown in Fig. 4. This behavior motivates the use of broadband sensors in the 77 GHz band to produce balanced images.

4.3 Maximum Velocity

According to (Reichthalhammer 2012), the maximum distance between sampling points is

$$\Delta x_{\max} = \frac{\lambda}{4 \cdot \sin(\gamma_{az}\ /\ 2)}. \tag{5}$$

With constant PRI, the sensor can feed a SAR algorithm at a maximum velocity of

$$v_{\max} = \frac{\Delta x_{\max}}{PRI}. \tag{6}$$

Table 2 shows the maximum velocity for all frequency bands.

Figure 3 shows the maximum velocity over beam width with a marker for a sensor with 80° beam width.

5 Evaluation Environment

A simulation and measurement environment has been implemented in MATLAB for evaluation purposes. The basic back projection algorithm is based on (Gorham and Moore 2010). Point-target simulations can be implemented by applying the

Table 2 Maximum velocity for SAR calculation for an assumed PRI of 150 µs

Radar operating frequency band (GHz)	Maximum sampling distance (mm)	Maximum velocity (km/h)
24.05–24.25	4.8	116
76–77	1.5	37
77–81	1.5	35

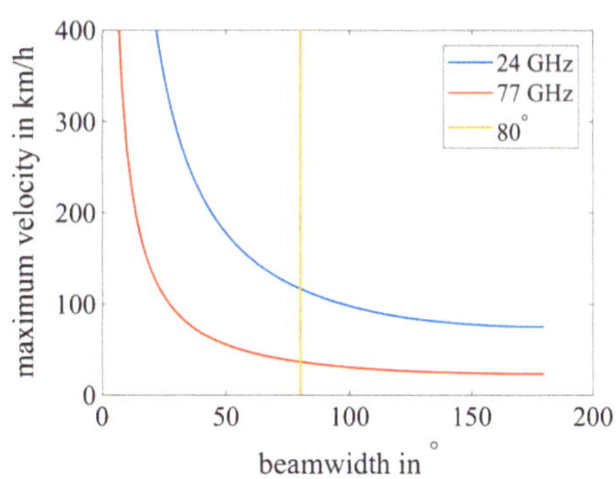

Fig. 3 Maximum velocity over beam width for SAR with an assumed PRI of 150 µs

Frequency Modulated Continuous Wave (FMCW) intermediate-frequency signal as derived in (Meta 2006):

$$s_{\mathrm{if}}(t,\tau) = A \cdot \exp\left(j2\pi\left(f_c\tau + \alpha t\tau - \frac{1}{2}\alpha\tau^2\right)\right), \tag{7}$$

where

$$\tau = \frac{2}{c_0} \cdot \sqrt{(\Delta x)^2 + (\Delta y)^2 + (\Delta z)^2} \tag{8}$$

is the double time delay between the sensor and the point target in three-dimensional Cartesian coordinates with the following parameters:

- A: Direction- and distance-dependent amplitude. A suitable angular reflection pattern can be used for directing objects.
- f_c: Carrier frequency of the sensor used, e.g., 24 GHz or 77 GHz.
- c_0: Free-space velocity of light, approximately 3×8 m/s.
- α: Frequency sweep rate, the slope of the FMCW system's frequency variation.

For surface targets, blocks of point targets can be used. Then Eq. 7 has to be solved and summed coherently for any point-target time delay τ in the point block (PB),

$$s_{\mathrm{if,PB}}(t,\tau) = \sum\nolimits_{\tau} s_{\mathrm{if}}(t,\tau_{\mathrm{i}}), \tag{9}$$

Alternatively, range plots simulated with a ray tracing tool can be used.

The complete SAR data-processing sequence is as follows:

- Simulation/Measurement of intermediate-frequency signal, as shown in Eq. 7.
- Calculation of range signal via Fast Fourier Transform (FFT).
- Mapping of trajectory and radar measurement, typically via an interpolation of the trajectory on the radar time stamps.
- Calculation of SAR image, as shown in Sect. 3.
- Calculation of the absolute value of any pixel in the SAR image.

6 Evaluation of Automotive Relevant SAR Properties

To evaluate SAR systems with series automotive sensors, further differences towards systems with special hardware are illuminated. A single-point target, as shown in Figs. 4 and 5, will serve as a reference. A comparison between simulation and measurement is displayed here. Figure 4 shows the simulation of a corner reflector. The reflector's reflectivity was simulated in an electromagnetic simulation software tool and applied to the complex amplitude, A, in Eq. 7. For the measurement in Fig. 5, a corresponding corner reflector, as shown in the background in

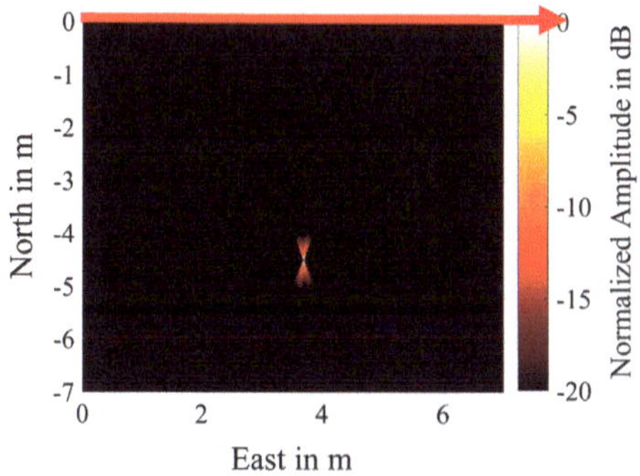

Fig. 4 Simulation of a point target

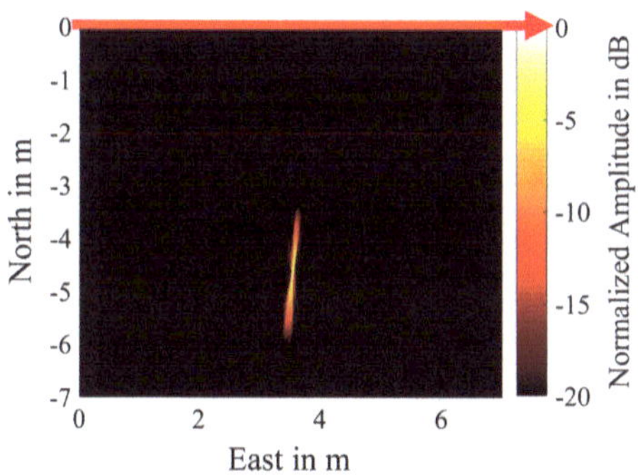

Fig. 5 Measurement of a point target

Fig. 1, was recorded. The SAR image shows the top view of the illuminated scene. The North axis represents the side-looking direction. The East axis represents the driving direction with the trajectory printed in red.

6.1 Incorrect Trajectory Measurement

Since high-performance odometry sensors are unavailable in series cars, an easy alternative approach is to use onboard information such as that from rotary encoders and a gyroscope, although doing so will definitely degrade trajectory quality. An error is introduced into the measured trajectory of the reference images in Figs. 4 and 5 to analyze the effect on the SAR image.

The greatest impact is achieved with lateral deflection (cross driving direction), as it effects phase variation the most. A sinusoidal error with a 2 cm peak-to-peak amplitude and a 1.5 m wavelength was used for the simulation. This model represents noise on the gyroscope data. The error leads to defocusing of the SAR image, which can be seen in Fig. 6 for the simulation and in Fig. 7 for the measurement. The green line shows the true trajectory; the red line shows the distorted

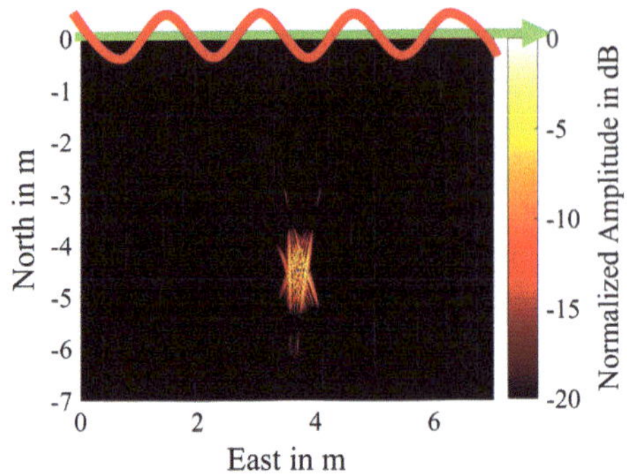

Fig. 6 Simulation: SAR image defocusing due to incorrect trajectory

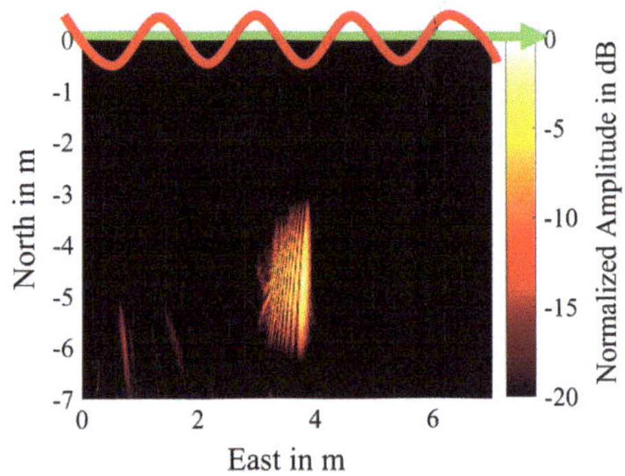

Fig. 7 Measurement: SAR image defocusing due to incorrect trajectory

trajectory used for SAR processing. To improve visualization, the amplitude is not to scale. Height and driving-direction errors do not corrupt the phase to the same degree as in lateral direction and are disregarded here.

The integration of an offset in gyroscopes and rotary encoders produces drifting errors. This effects the image in two different ways: displacement of the detected targets, and focus degradation, the latter of which is negligible. Drifting clocks in the rotary encoder and the radar system engender similar behavior.

6.2 *Time-Based Sampling*

As shown in Sect. 3 with a series sensor, a spatial trigger cannot initiate an individual radar measurement, e.g., one measurement every 4 mm. Radar measurements have to be accepted as the sensor provides them, e.g. with a PRI of 150 μs. So radar measurements have to be localized via temporal mapping, which is done by interpolating the radar and odometry time stamps. Besides introducing an additional positioning error, this may impact spatial sampling distance at high velocities. As derived in (Gerbl 2007), a maximum spatial sampling distance must

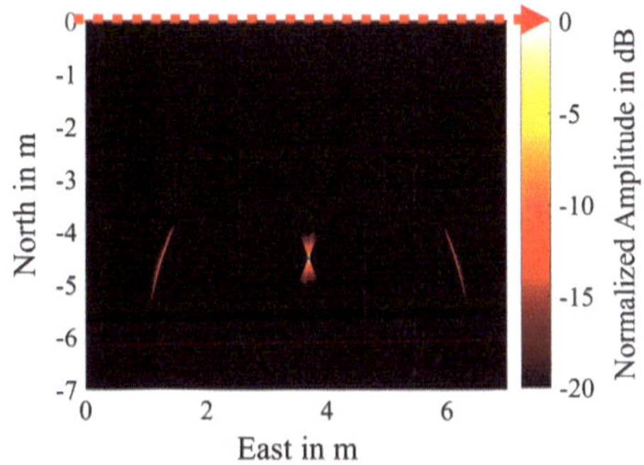

Fig. 8 Simulation: aliasing due to insufficient azimuth sampling

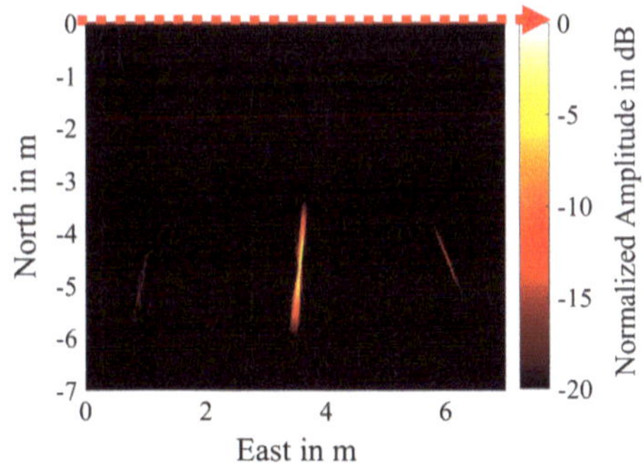

Fig. 9 Measurement: aliasing due to insufficient azimuth sampling

not be exceeded during SAR measurement because aliasing effects (ambiguities) analogous to those in FFT occur. Figure 8 shows effects on the resulting SAR image for the simulation and Fig. 9 shows those for the measurement. Here the azimuth sampling distance was increased from 4 to 12 mm in both cases. This is indicated by the dotted trajectory line in the corresponding figures.

Consequently, it must be guaranteed, that the maximum velocity of the specific sensor, as derived in Sect. 3, is not exceeded.

7 Simulation and Measurement

The described system is tested by simulating and measuring the parking-lot scenario shown in Fig. 10. Four parked cars, (1) to (4), and an interrupted curbstone, (5) in the background, fall within the area under consideration. The discontinuity is located just left below label (5). The recording vehicle is marked with (6). The red arrow indicates the sensor's trajectory.

Fig. 10 Parking-lot scenario

7.1 Measurement

For the measurement a standard automotive series sensor has been modified to output raw Analog-to-Digital Converter (ADC) data. Only one of the eight available channels is evaluated, as no beam forming is done here. This produces the broadest possible beam width and along with that highest azimuth resolution. The raw data from a side-looking radar has been recorded and input to the MATLAB system. Odometry data was generated by rotary encoder and gyroscope measurements. Global Positioning System (GPS) time stamps are used to map between radar and odometry data. Figure 11 shows the results.

The image shows the top view with azimuth (East) and range (North) direction. The color code represents the corresponding pixel's phase-corrected amplitude and could be interpreted as an occupancy probability. As the radar signal does not

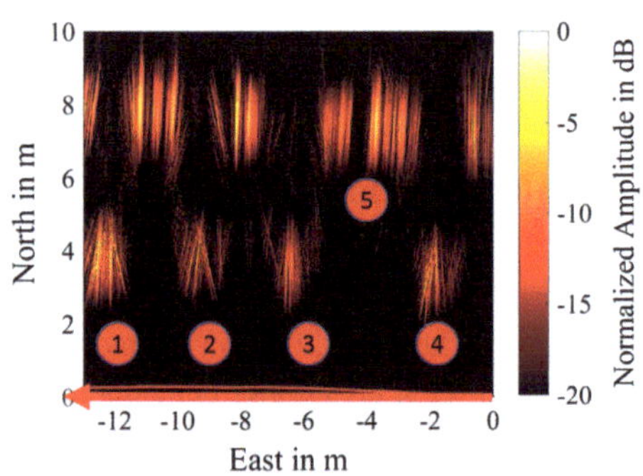

Fig. 11 Measurement of the parking-lot scenario

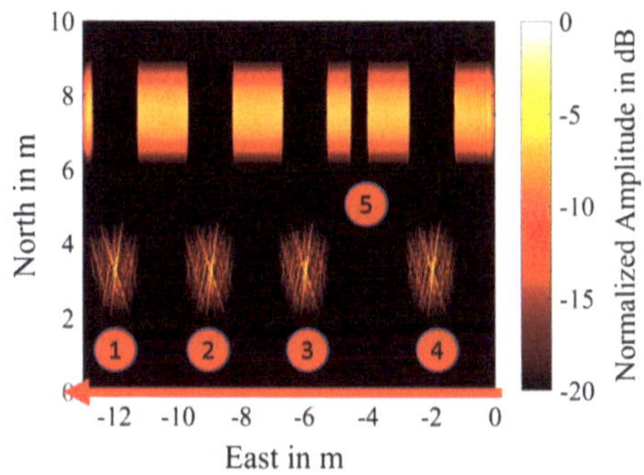

Fig. 12 Simulation of the parking-lot scenario

penetrate the cars' metal surfaces, only the impact of the first hit is measured. Objects located behind them are shadowed and undetected. Range resolution is limited to 1 m as derived in Sect. 4.2, which explains why the image is blurred in the range direction rather than appearing as a slim contour line. The scale is given in decibels relative to the brightest pixel. The curbstone is clearly visible, as it forms a dihedral corner reflector together with the ground. The discontinuity just below the (5) marker is also visible. In contrast to the curbstone, the cars' rounded corners reflect most of the signal energy away from the sensor. The latter's surface structures reflect only a small amount back to the sensor. This effect is expressed by the fact that the high power reflection of cars seems to be narrower in the SAR image than they are in reality. So a large dynamic range is necessary to detect the corresponding areas of parked cars.

7.2 *Simulation*

For the simulation, a grid of point targets, representing the cars and curbstone as shown in Fig. 10, is implemented as described in Sect. 5. The distance between two points is 5 mm in the x, y, and z direction. All point targets are omni-directional reflectors with equal amplitude, so no material characteristics are considered. Only first hit points are implemented to represent a shadowing effect. The results can be seen in Fig. 12.

8 Conclusion

In this paper, we have presented a possibility for increasing the azimuth resolution of automotive radar sensors. SAR processing seems to be feasible with current series sensors. The azimuth resolution can be increased from several decimeters to a few centimeters. With that, an environment map can be created using only current hardware and additional signal processing. Measurements with a 24 GHz sensor were presented to show the system's potential. The SAR algorithm is applied after the actual radar measurement, and with that independent of the sensor. So 77 GHz sensors are planned for future measurements to benefit from the broader bandwidth and associated range resolution.

Acknowledgements The research leading to these results was conducted during a cooperation project between perisens GmbH, Continental, Chassis and Safety Division, ADAS Business Unit and Volkswagen Aktiengesellschaft.

References

Cumming I, Wong F (2005) Digital Processing of Synthetic Aperture Radar Data: Algorithms and Implementation, ser. Artech House remote sensing library. Artech House, no. Bd. 1

Gerbl F (2008) Evaluation of wide beam, short-range synthetic aperture radar imaging: Techn Univ Diss–München, 2007. Berlin: Logos-Verl

Gorham L A, Moore L J (2010) SAR image formation toolbox for MATLAB, pp. 769 906–769 906–13, http://dx.doi.org/10.1117/12.855375

Iqbal H, Sajjad M, Mueller M, Waldschmidt C (2015) SAR imaging in an automotive scenario. In: 2015 IEEE 15th Mediterranean microwave symposium (MMS), pp. 1–4

Klausing H, Holpp W (1999) Radar mit realer und synthetischer Apertur: Konzeption und Realisierung. De Gruyter

Meta A (2006) Signal processing of FMCW Synthetic Aperture Radar data. [S.l.]: [s.n.]

Mure-Dubois J, Vincent F, Bonacci D, Sonar and radar SAR processing for parking lot detection. In: 2011 12th International radar symposium (IRS), pp. 471–476

Reichthalhammer T (2012) Ein Radar mit synthetischer Apertur für den Nahbereich: Techn Univ Diss–München, 1st edn, ser. Ingenieurwissenschaften. München: Dr. Hut

Skolnik M (2008) Radar Handbook, Third Edition, ser. Electronics electrical engineering, McGraw-Hill Education

Winner H, Hakuli S, Wolf G (2009) Handbuch Fahrerassistenzsysteme: Grundlagen, Komponenten und Systeme für aktive Sicherheit und Komfort: mit 550 Abbildungen und 45 Tabellen, ser. ATZ-MTZ-Fachbuch. Vieweg + Teubner

Wu H, Zwick T (2009) Automotive SAR for parking lot detection. In: 2009 German microwave conference, pp. 1–8

Wu H, Zwirello L, Li X, Reichardt L, Zwick T (2011) Motion compensation with one-axis gyroscope and two-axis accelerometer for automotive sar. In: 2011 German microwave conference, pp. 1–4

SPAD-Based Flash Lidar with High Background Light Suppression

Olaf M. Schrey, Maik Beer, Werner Brockherde and Bedrich J. Hosticka

Abstract In this contribution, we present the concept of a 4×128 pixel line sensor for direct time-of-flight measurement based on single-photon avalanche diodes (SPAD) fabricated in a high-voltage automotive 0.35 μm CMOS process. An in-pixel time-to-digital converter with a resolution of 312.5 ps determines the arrival of photons reflected from targets in the area of view. Since we are employing a so-called first photon approach, there are no dead-time effects. In addition, our approach uses a variable photon coincidence detection to suppress effects of ambient illumination. As a test vehicle we have implemented a 1×80 pixel CMOS SPAD line sensor and characterized it.

Keywords Lidar · Time-of-Flight · CMOS-Chip

1 Introduction

Today's advanced driver assistance systems and autonomous driving systems require fast and reliable sensing and monitoring of the vehicle environment in order to detect obstacles (Ackerman 2016). Various technologies can be employed to provide sensory information for this purpose, such as radar or computer vision. One

O.M. Schrey (✉) · M. Beer · W. Brockherde
Fraunhofer Institute for Microelectronic Circuits and Systems (IMS), Finkenstr. 61, 47057 Duisburg, Germany
e-mail: olaf.schrey@ims.fraunhofer.de

M. Beer
e-mail: maik.beer@ims.fraunhofer.de

W. Brockherde
e-mail: werner.brockherde@ims.fraunhofer.de

B.J. Hosticka
Department of Electronic Components and Circuits, University of Duisburg-Essen, Bismarckstr. 81, 47057 Duisburg, Germany
e-mail: bedrich.hosticka@ims.fraunhofer.de

C. Zachäus et al. (eds.), *Advanced Microsystems for Automotive Applications 2017*,
Lecture Notes in Mobility, DOI 10.1007/978-3-319-66972-4_4

of the most promising technologies is the lidar, which stands for light detection and ranging. Most of the today's lidar sensors employ laser scanners to scan the target scene. Since laser scanners are bulky and highly costly due to the required high-precision mechanical parts, affordable solid-state flash lidar sensors are the key factor when it comes to mass market introduction of lidar systems. With flash lidar systems, the entire scene is illuminated with a single laser beam shot without requiring moving parts and thus considerably reducing system costs. As photo sensors, single-photon avalanche diodes (SPADs) are the optimum choice since they are very fast, achieve a very high sensitivity, and can be fabricated in standard CMOS processes (Bronzi et al. 2012; Beer et al. 2016). Two principal techniques for distance measurement based on time-of-flight (ToF) can be distinguished: indirect and direct ToF (Bronzi et al. 2012; Brockherde 2001; Spickermann et al. 2011). While the indirect method relies on photon counting, the direct approach measures the elapsed time between the emission and reception of a narrow laser pulse reflected from a distant target. However, for the timing measurement a time-to-digital converter (TDC) is required: its precision greatly affects the accuracy of the distance measurement. In addition, the reflected laser signal typically is buried in ambient illumination signal and has to be recovered.

2 Sensor Principle

In this paper, we present a 4×128 pixel line sensor design for direct ToF measurement based on SPADs fabricated in a high-voltage automotive 0.35 μm CMOS process. Here, the SPADs are perfectly suitable for direct ToF imaging since they exhibit a timing resolution in the picosecond range and generate directly digital output signals. Nevertheless, the SPADs require so-called quenching and reset circuits and exhibit dead-time effects (Bronzi et al. 2012; Beer et al. 2016, 2017; Yu and Fessler 2000). While the quenching circuit tends to reduce the fill factor and can be combat by using simple circuits, the reset circuit causes a saturation for high photon flux rates.

In the line sensor, each pixel contains a quenching and reset circuit and a TDC for precise time measurement. In our approach, the TDC determines the arrival time of only the first photon—though we have to recall that in reality owing to the quantum efficiency, the SPAD "fires" only when the first electron is photo generated—at each pixel separately during a measurement period, but in parallel in all pixels. All that is needed to determine a target distance for a single pixel, is the arrival time of the first photon (or actually, the first photo electron for that matter) reflected from the target. Note, that this ToF approach is thus not affected by the dead-time effects, typically for SPADs, although it requires a sufficient number of repeated measurements. A very useful secondary effect is that with the "first photon" method the detector will never saturate, even with very strong reflected laser light.

Needless to say that photon arrival times are subject to photon statistics. If we consider only the reception of reflected laser pulses, then the arrival time can be determined by filling the time bins over multiple pulses in a histogram and applying software algorithms to process adequately the photon statistics for distance computation. However, since we are also recording photons stemming from the ambient light, we had to develop a suitable suppression technique for ambient illumination. The influence of ambient light on the distance measurement can be greatly reduced applying coincidence detection: this approach utilizes the occurrence and detection of temporal and/or spatially correlated photons (Hayat et al. 1999). The coincidence detection enables separation of photons generated by reflected laser pulses. Their time of travel corresponds linearly with the target distance. Photons generated by ambient light exhibit a random characteristic and can be separated from the active laser light by employing coincidence statistics. After the separation of the ambient signal from the reflected laser signal, we can readily determine the reception time of the laser signal and thus calculate the target distance. Hence, the coincidence detection results in a substantial lidar performance improvement of the system. The sensor presented here employs variable photon coincidence calculation from four single SPADs in each pixel which offers a much better performance when compared to fixed coincidence calculation (Niclass et al. 2013). By adjusting the time and depth of the coincidence detection, a robust distance measurement for varying ambient lighting conditions is realized. By suppressing ambient illumination, the range and accuracy of a distance measurement system are greatly increased.

3 Technology and Measurements

As a first test vehicle we have implemented a 1×80 pixel linear SPAD sensor, which was fabricated in a high-voltage automotive 0.35 μm CMOS process (see the chip photomicrograph in Fig. 1). This process yields SPADs with very low dark count rate and low after-pulsing rates.

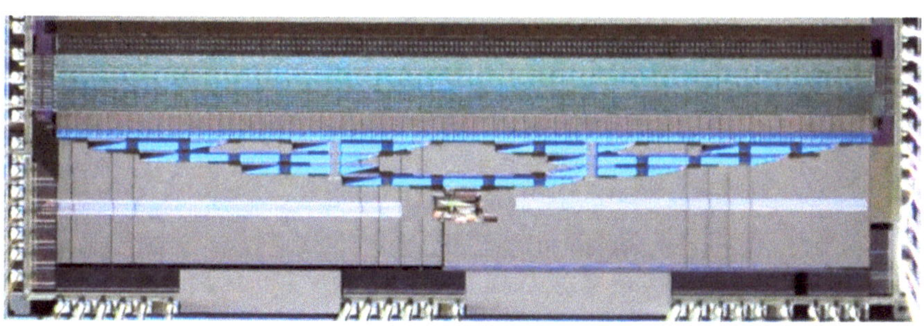

Fig. 1 Chip photomicrograph

The chip area is 46.45 mm^2, while the pixel pitch is 100 µm and the fill factor is 19%. The chip dissipates 1.05 W at 3.3 V power supply voltage and operates at 200 MHz clock frequency, although the SPADs require 24 V. The TDC range is 1280 ns and its resolution is 312.5 ps which corresponds to a distance resolution of 4.69 cm. The sensor features variable photon coincidence circuitry with variable coincidence time and depth adjustment. The maximum coincidence depth level is 4. The readout clock rate is 20 MHz, so that the total readout time is 4 µs. For single distance measurements we used 400 laser pulses. These measurements yielded a detection range of 12 m at an ambient illumination of 90 klux, while the field-of-view (FOV) was 0.46° × 36.8°. The laser FOV was 2° × 40°. We used a commercially available laser exhibiting a laser power of 75 W, 20 ns pulse width, and 10 kHz laser repetition rate, while the wavelength was 905 nm. The temperature drift of the measured distance amounted to 2.2 mm/K. Some measurements are shown in Figs. 2, 3, and 4.

Currently we are developing a 4 × 128 pixel line sensor based on the same principle, which has an increased detection range of 30 m with a 40° horizontal FOV at an ambient illuminance of 60 klux for a white target with 80% reflectance. The laser exhibits again a laser power of 75 W, 20 ns pulse width, and 10 kHz laser rate, while the wavelength is 905 nm. The sensor floorplan and layout allows the development of a custom multi-line sensor design with, e.g., four lines with custom spacing (Fig. 5). This enables multi-line flash detection of the car exterior with high

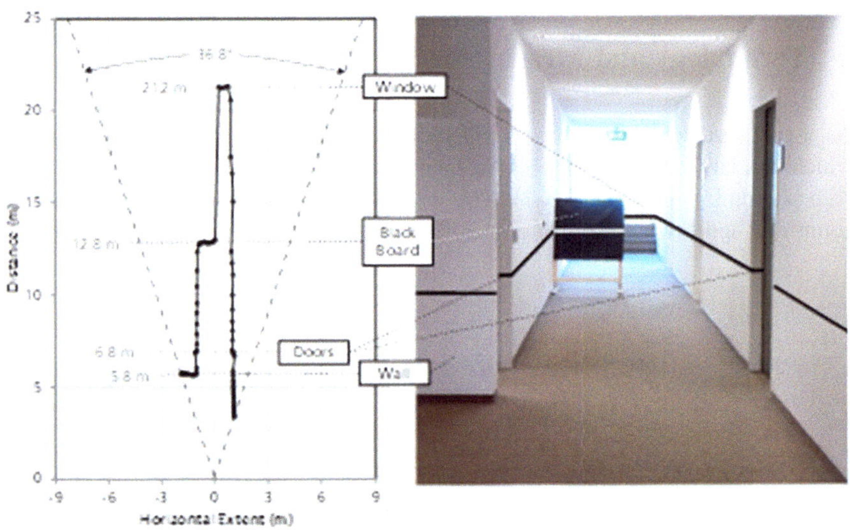

Fig. 2 Indoor measurement of the 1 × 80 linear sensor. On the right hand side the target scene is shown. On the left side the dashed lines mark the sensor FOV and the solid line indicates the measured distance

Fig. 3 Setup of the sensor range measurement at high ambient illumination. The target (e.g. white paper with 80% reflectance) is placed in increasing distance in front of the laser source and sensor. In this example an ambient illuminance of 90 klux was measured at the target surface

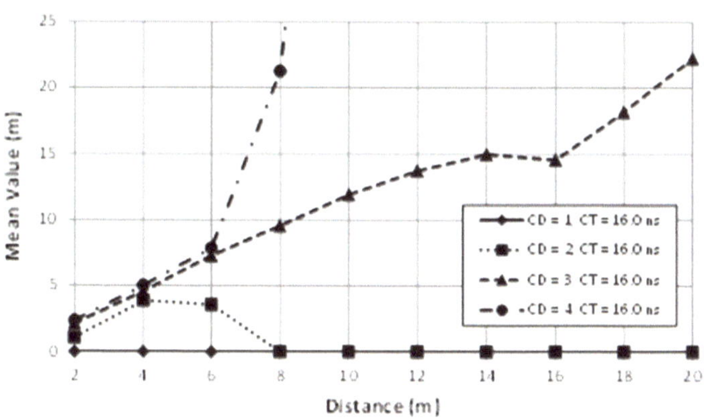

Fig. 4 Results of the distance measurement for a white Lambertian target shown in Fig. 3 at an ambient illumination of 90 klux for coincidence depth from 1 to 4 levels and coincidence time of 16 ns. Each data point has been obtained from 100 distance measurements with 400 laser pulses each. Here the coincidence depth of 3 yields the best result at a range of 12 m

lateral resolution in combination with an extended vertical detection range. Typical application scenarios would cover 40° … 80° horizontal and up to 20° vertical field-of-view (see Fig. 6).

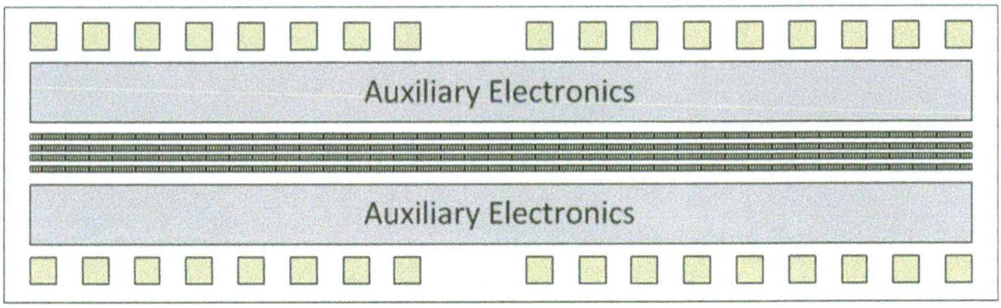

Fig. 5 Design example of a SPAD-based multi-line flash lidar sensor in a standard CMOS process

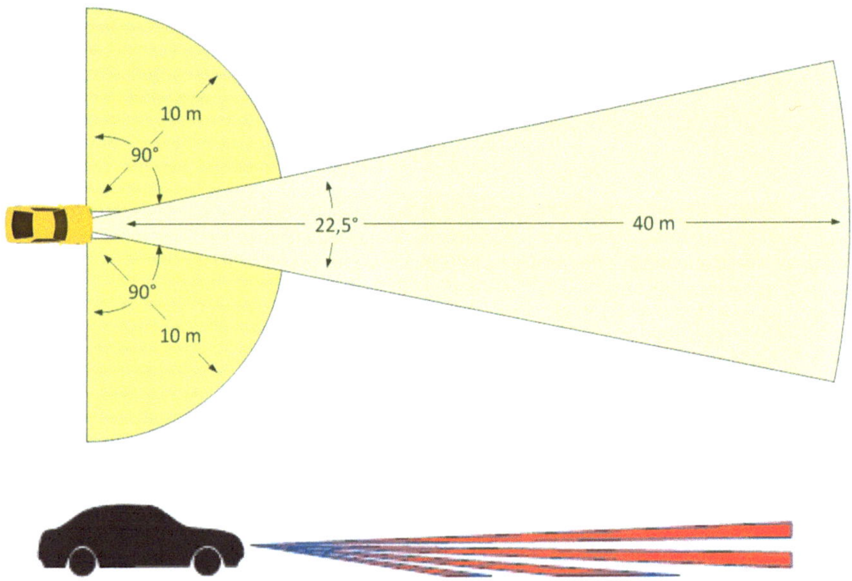

Fig. 6 Automotive application of the SPAD-based multi-line flash lidar sensor

4 Summary

We have presented a novel SPAD multi-line sensor for automotive flash lidar applications fabricated in a standard CMOS process. It is based on direct time-of-flight approach and uses the first photon detection method, so that it does not suffer from dead-time effects and allows for very high dynamic range without saturation effects. This approach, however, requires statistical evaluation of photon arrival times. The sensor employs variable coincidence time and depth detection to suppress effects of ambient light. Initial tests based on a single SPAD line sensor proved clearly the feasibility of this approach, even under very high ambient illumination.

References

Ackerman E (2016) Lidar that will make self-driving cars affordable [News]. IEEE Spectr 53 (10):14

Bronzi D, Villa F, Bellisai S, Markovic B, Tisa S, Tosi A, Zappa F, Weyers S, Durini D, Brockherde W, Paschen U (2012) Low-noise and large-area CMOS SPADs with timing response free from slow tails. In: Proceedings of the European solid-state device research conference (ESSDERC), pp 230–233

Beer M, Hosticka B J, Kokozinski R, (2016) SPAD-based 3D sensors for high ambient illumination. In: Conference on Ph.D. research in microelectronics and electronics (PRIME), pp. 1–4

Jeremias R, Brockherde, W, Doemens G, Hosticka B, Listl L, Mengel P (2001) A CMOS photosensor array for 3D imaging using pulsed laser, digest of technical papers. In: IEEE international solid-state circuits conference, pp. 252–253

Spickermann A, Durini D, Süss A, Ulfig W, Brockherde W, Hosticka B J, Schwoppe S, Grabmair A (2011) CMOS 3D image sensor based on pulse modulated TOF principle and intrinsic LDPD pixels, In: Proceedings of the European solid-state circuit conference (ESSCIRC), pp 111–114

Yu DF, Fessler JA (2000) Mean and variance of single photon counting with deadtime. Phys Med Biol 45(7):2043–2056

Beer M, Schrey O, Hosticka B J, Kokozinski R (2017) Dead time effects in the indirect time-of-flight measurement with SPADs, In: IEEE international symposium on circuits and systems, vol 1168, session A3L-G

Hayat MM, Torres SN, Pedrotti LM (1999) Theory of photon coincidence statistics in photon-correlated beams. Optics Commun 169(1–6):275–287

Niclass C, Soga M, Matsubara H, Kato S, Kagami M (2013) A 100-m range 10-frame/s 340 × 96-pixel time-of-flight depth sensor in 0.18 μm CMOS. IEEE J Solid-State Circuits 48 (2):559–572

Part II
Driver Assistance and Vehicle Automation

Enabling Robust Localization for Automated Guided Carts in Dynamic Environments

Christoph Hansen and Kay Fuerstenberg

Abstract The range of applications for autonomous guided carts (AGC) is increasingly growing. Especially in industrial environments ensuring high safety standards in combination with high availability and flexibility are major requirements. For this reason, knowledge about its own position in the environments becomes particularly important. For AGC with low vehicle height localization approaches based on contour observations are widespread. However, in over-time-changing environments the robustness of these techniques is limited. This paper proposes an approach for updating the underlying map in real time during operation. This map update allows for a long-term robust localization. The proposed approach is evaluated for a dynamic test scenario using a cellular transport vehicle.

Keywords Map update · Dynamic environment · Localization · Pose estimation · Long term · Robust · Accuracy evaluation · Autonomous guided vehicle · AGV · Autonomous guided cart · AGC · Industrial applications

1 Introduction

Mobile transport robots in the market, used for various logistic applications, differ considerably in their design, characteristics, and features. Besides automated forklift trucks, vehicles with a significantly lower height became successfully established in the market. For those autonomous guided carts a trend towards increased flexibility has become apparent.

A prerequisite to ensure flexible and also safe operation is the knowledge of the current position of the AGC within its environment. Established localization

C. Hansen (✉)
SICK AG, Merkurring 41, 22143 Hamburg, Germany
e-mail: christoph.hansen@sick.de

K. Fuerstenberg
SICK AG, Erwin-Sick-Str. 1, 79183 Waldkirch, Germany
e-mail: kay.fuerstenberg@sick.de

C. Zachäus et al. (eds.), *Advanced Microsystems for Automotive Applications 2017*,
Lecture Notes in Mobility, DOI 10.1007/978-3-319-66972-4_5

technologies using artificial landmarks in the environment (e.g., reflectors) require uneconomical high installation effort and costs. Alternative localization approaches use contour observations of the environment, often using a laser range sensor, for estimating the position of the AGC in a previously created map. The map is commonly created using a SLAM algorithm (Beinschob and Reinke 2015), while for the localization in this map the Monte Carlo-Localization (MCL) algorithm is used in many applications (Gutmann et al. 1998; Dellaert et al. 1999) (Gustafsson et al 2002) (Kirsch et al 2002). Due to the low vehicle height, the contour-based localization approaches face the challenge of a predominant non-static environment, as many objects in logistic environments (e.g. pallets, goods, other vehicles and humans) move in the course of time (Fig. 1). These dynamic objects hamper a long-term reliable localization of an AGC. Over time, less information from the contour observation matches the static map and thus the uncertainty of the resulting MCL poses increases. Hence, in dynamic environments the MCL cannot guarantee a reliable localization result based on the previously created static map. Moreover, the high amount of non-static parts in the environment results in a need for an update of the static map.

In dynamic environments, two different types of non-static objects need to be differentiated for updating the digital map of the environment. High-dynamic objects are characterized by a relatively small time they stop at one position compared to the time they are moving. Typical examples are other vehicles and moving persons. In contrast, semi-dynamic objects stay at one position for a longer period. Here, typical examples are pallets or parking vehicles. Both types disturb the localization process when they leave their initial position. However, only semi-dynamic objects could potentially serve as supporting landmarks, as these objects stay at specific positions for a longer time. Hence, including their contour in the map used for localization improves the resulting vehicle pose estimation. Frequently including the contour of high-dynamic objects does not enhance the accuracy and robustness of the localization process, as the updated contours will be outdated immediately. Besides using an updated map for localizing a mobile transport robot, the updated information on the environment can also be used for a

Fig. 1 Estimating the position of a free navigating AGV with limited height based on distance measurements is affected by changes in the environment (boxes in the open space)

more sophisticated navigation and path planning process. In this case, the position of both types of non-static objects could be of interest for global and local path adaption (Kleiner et al 2011).

In the course of this paper, we propose a localization method ensuring a long time robust and reliable localization in dynamic environments. The Monte Carlo-Localization with Map Update (MCL/MU) approach is structured in four iterative phases, starting with a localization phase, followed by a control, an update and a fusion phase. The approach is designed for AGC in industrial environments with particularly high standards for transport reliability and availability. Therefore, the required technical specifications of a maximum position error of 0.1 m and an orientation error of less than 1.5 degree need to be realized with limited computational power and memory, but with real-time capability. As we concentrate on the localization the proposed map update approach is optimized for updating the contour of semi-dynamic objects. However, the design allows for a more sensitive update of high-dynamic objects by using a different parameter set.

The proposed MCL/MU algorithm is evaluated in a dynamic scenario at the Fraunhofer Institute for Material Flow and Logistics (IML) using a Cellular Transport Vehicle (CTV), equipped with two SICK S300 safety laser range finders. An additionally mounted SICK NAV350 laser navigation sensor provides a reference position, by observing reflectors in the environment.

2 Related Work

Current approaches for localization of an AGV in dynamic environments follow three different basic concepts. The first fundamental idea is an outlier filter for incoming sensor readings. Using only measurement readings on static objects prevents the localization process from being disturbed by dynamic objects. However, as the number of dynamic objects increases, the remaining information decreases. Having too few information hinders an accurate localization. Fox et al. (Fox 2003) presents two different filter techniques to detect single measurements in a scan caused by highly dynamic objects, based on an entropy value for a scan and comparison between expected and measured range. Schulz et al. (2003) proposes a filter which analyses local minima in a complete range measurement scan. Thereby particularly dynamic objects in front of static walls can be detects. Fujii et al. (Fujii et al 2015) also focus on the problem having too few information left for a reliable localization. Their approach detects localization failures and falls back to a simple odometry based localization for these cases.

The second already discovered idea to estimate an accurate position in dynamic environments is using one or multiple dynamic maps in addition to an unchanged static map. Thereby these methods provide additional information on dynamic objects for the localization process. However, the separate steps for localization in a static and a dynamic map requires additional processing time and resources. Valencia et al. (Valencia et al 2014) propose a localization method for dynamic environments based on a static and a continuously updated map for dynamic objects in the environment. First, a localization is performed in the unchanged static map using the well-proven MCL algorithm. For measurement readings not matching to an object in the static map, the dynamic map is used for calculating the weight of the respective particle. Together the readings on static and dynamic objects result in a combined weight for each particle. For continuously updating the dynamic map the probability of the existence for a dynamic object is calculated. An actual update is only implemented in case of high certainty for the currently estimated pose. Similarly also Meyer-Delius (Meyer-Delius 2011) implements an approach combining a localization in a static and supporting dynamic maps. A calculated outlier ratio provides information on the ratio of measurement readings not matching the static map and the total number of readings in a scan. For a small outlier ratio the unchanged static map is used for localization. In case of a higher outlier ratio the proposed approach checks if a temporary, local map is available and whether this map still reflects the current setup of the dynamic environment, i.e. whether the outlier ratio is reduced sufficiently. If there is such a temporary, local map, it is used for the localization process. Otherwise, a new temporary map for the current setup of the local, dynamic environment is created.

The third fundamental idea is using a single map which is periodically updated to the current setup of the dynamic environment. Thus, each localization step can be done using a single map. However, this bears the risk of an error propagation for the long-term localization. Mistakes in the map update, caused by inaccuracies of either the localization or the sensor measurement, affect following localization results and thereby also subsequent map update steps. Hence, great and time-consuming effort is required in order to ensure an accurate map update. Tipdaldi et al. (2013) propose a dynamic map update inspired by methods for simultaneous localization and mapping (SLAM) (Fernandez-Madri gal 2012). For each particle a separate map is updated for every incoming scan. The update is based on the Hidden Markov Model (Mongillo and Deneve 2008) and a calculation of transition probabilities for the state of each cell in the map. Carrying and updating a separate map for each particle results in high computational complexity. The described existing approaches do either not result in reliable, accurate, long-term localization or they are not capable to calculate poses in real time. Thus they are not applicable in large industrial environment, such as warehouse application with limited resources. To overcome these issues our approach reduces the complexity of the dynamic update of a single map used for localization. The core of our approach is a combination of a pose accuracy estimation and a subsequent map update.

3 The MCL/MU Approach

Our approach for localization of a mobile robot in dynamic environments uses a single, regularly updated map, reflecting the current setup of the environment. This allows separating the localization from the map update process and thus use any proven localization method, such as the MCL algorithm. Furthermore, using a single map for the actual localization process limits the required computation time. In order to prevent from inaccurate map updates and resulting localization error propagation the approach first estimates the accuracy of the determined pose, before allowing an actual map update. Optionally an unchanged static map can be used for prohibiting a map update for static pixels. The approach is designed for AGC in warehouse environments with particularly high standards for safety, availability, and transport reliability. Derived from the field of application, the approach aims to guarantee a long-term robust localization at a maximum position error of 0.1 m. Consequently, the map update should avoid errors to the greatest possible extend. This in turn requires the error in orientation for the estimated position to be less than 1.5 degree. Furthermore, in order to be applicable in real world industrial and logistic application the implementation needs to be real-time capable.

The proposed approach is structured in four iterative phases. In a first phase a localization in the current map is performed. Here, a MCL algorithm is used which is extended by calculating meta-information used in the following control phase. The map update control phase evaluates the accuracy of the estimated vehicle pose. For sufficiently high accuracy estimates a dynamic map update is triggered in the third phase. Finally, in a fourth step the identified changes in the map are collected, fused and integrated in the MCL algorithm. Hereafter, the three iterative phases following the localization step are presented in more detail.

3.1 Map Update Control

The deviation between the pose estimated in the localization phase and the actual position of the mobile robot in its environment cannot be quantified directly, as the true position is unknown. Thus, in order to make a statement on the accuracy of the pose resulting from the localization, other explaining criteria are required, which can be directly derived from the MCL algorithm. Several criteria were tested and analyzed regarding their correlation with the actual position error and also the correlation to other criteria. Eventually, two criteria were found to provide significant indication on the position error without correlating with each other. The first criterion (OR_{best}) describes the outlier ratio for the respective best particle pose, i.e., the ratio between measurement reading matching the current map and the total number of readings in the scan. (OR_{best}) has a correlation coefficient with the actual position error of 0.69. The second criterion (PDbest) describes the average particle distance to the respective pose of the best particle, i.e., the scattering of the used particle poses. Here, the particle distance is defined as an average between position and orientation difference. The resulting coefficient error for PD_{best} and the actual position error is at 0.81, while the correlation to OR_{best} turns out to be 0.08.

$$OR_{best} = \frac{outlier}{totalbeams} \tag{1}$$

$$PD_{best} = \sum_{i=0}^{n} \frac{dist_{i,best}}{n} \tag{2}$$

To take a decision on updating a map, a robust binary selector is required differentiating between a successful or failed localization. Having found two almost independent criteria both indicating the accuracy of the localization, a logistic regression model (Pampel 2000; Behnke 2014) provides an appropriate tool for combining the two criteria into a binary selector. In order to derive the logistic regression model, Data from a training set is divided into a group of successful localization and a second group of failed localization. The classification is done based on the deviation of the estimated pose and a reference pose determined using installed reflectors in the environment (Fig. 2). The resulting logistic regression model (Eq. 3) allows differentiating between successful and failed localization with an average certainty of 90%.

$$logit(p) = -4.457 + 8.225 * OR_{best} + 1.124 * PD_{best} \tag{3}$$

Shifting the regression model parallel downwards enables recognizing nearly all failed localization at the cost of classifying more successful localization results as failures. However, for the intended application it is crucial to detect all failed localization results in order to hinder map update failures.

3.2 Map Update and Map Update Fusion

For high certainty on the currently determined pose of the mobile robot, an update for the map according to the current setup of the environment is performed. The map update approach is based on the proposed approach of Tipdaldi et al. (2013) and puts special emphasis on resource efficiency. This update phase aims to keep the occupancy state of dynamic pixels up to date. This information then contributes

Fig. 2 Logistic regression model

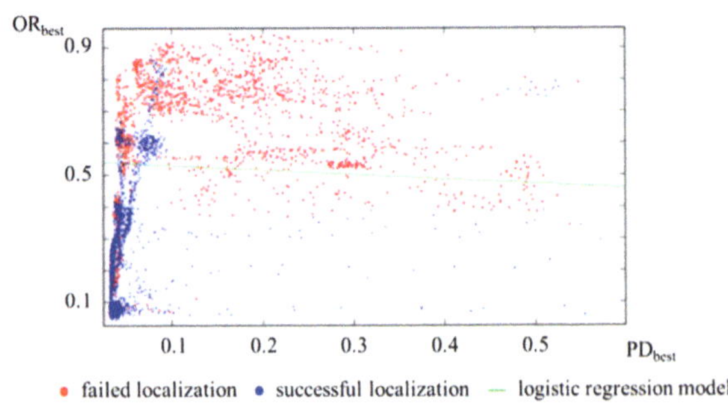

to reliable results in the next localization phase. However, for this purpose only semi-dynamic objects provide additional value. Therefore an update for dynamic objects, such as other moving vehicles or persons, is not necessary. In order to guarantee a precise update and avoid an error propagation, a small delay for changing the occupancy state of a pixel is desirable. Thereby fast oscillating occupancy states for a pixels and thus potential errors are avoided. The dynamic map updated implemented for the proposed approach is based on the calculation of state transition probabilities $p(c_{t+1}|c_t)$ for each pixel in the map (Eq. 7). These describe the probability for the upcoming occupancy state of a pixel c_{t+1}, with knowledge of the current pixel state c_t. For each pixel an occupancy state probability vector Q_t (Eq. 5) is updated for the next time step Q_{t+1} (Eq. 6). In order to limit the required time for computation this update is only performed for currently observed pixels. The actual update results from a multiplication of the current occupancy state probability vector Q_t, the state transition probability Matrix A_c, an observation probability matrix B_z and a normalization factor η (Eq. 4). The observation probability matrix B_z reflects uncertainties in the sensor measurements (Eq. 8) and the normalization factor η ensures that the sum for the probabilities of a free and an occupied pixel state equals one at all times.

$$Q_{t+1} = Q_t * A_c * B_z * \eta \tag{4}$$

$$Q_t = \left[p(c_t = occ|z_{1:t}) \quad p(c_t = free|z_{1:t}) \right] \tag{5}$$

$$Q_{t+1} = \left[p(c_{t+1} = occ|z_{1:t}) \quad p(c_{t+1} = free|z_{1:t}) \right] \tag{6}$$

$$A_c = \begin{bmatrix} p(c_{t+1} = occ|c_t = occ) & p(c_{t+1} = free|c_t = occ) \\ p(c_{t+1} = occ|c_t = free) & p(c_{t+1} = free|c_t = free) \end{bmatrix} \tag{7}$$

$$B_z = \begin{bmatrix} p(z|c = occ) & 0 \\ 0 & p(z|c = free) \end{bmatrix} \tag{8}$$

The occupancy probability matrix B_z depends only on the parameters of the laser rangefinder and the chosen resolution of the map. Hence, the respective matrices for an observation of an occupied cell and also an observation of a free cell can be specified once in the beginning. In the following, both can be used as a constant. In contrast, the state transition probability matrix A_c needs to be update for each iterative update phase. For this purpose an exponential moving average for each state transition probability is calculated separately. Thereby, more recent observations have a higher influence on the updated probabilities then older observations. Using an exponential moving average and updating the dynamic information only for pixels within the viewing range of the laser rangefinder allows for a reduction in computation time. An actual map update is performed if the occupancy probability exceeds or falls below a variable threshold. In the following fusion phase, changes are clustered in order to allow a time efficient update of the likelihood-field used for the MCL.

4 Evaluation

The proposed approach is evaluated in a dynamic scenario at the Fraunhofer Institute for Material Flow and Logistics (IML) using a Cellular Transport Vehicle (CTV). The CTV fleet at the IML consists of 50 vehicles, each equipped with two SICK S300 safety laser range finders, a differential drive and a load handling device to carry boxes in logistics environments with up to 40 kg (Kamagaew et al 2011). An additionally mounted SICK NAV350 laser navigation sensor in the center of the CTV (Fig. 3) provides a reference position. This reference position is obtained by observing reflectors in the environment and used for evaluating the accuracy of the estimated poses. When recording the test date for the dynamic scenario, the used CTV moved at a speed of 0.8 m per second. Controlled by a multi-agent system, a particle filter based localization was performed on the vehicle (Kirsch and Roehrig 2011) and all occurring sensor readings safety laser rangefinder, laser navigation sensor, odometry were logged on a second computer for offline evaluation.

The data of the dynamic scenario is recorded at the Fraunhofer IML research laboratory with an area of more than 1.000 square meters to evaluate coordination, localization a mapping algorithms with a focus on real-world logistic applications. A static map of the test environment is given in Fig. 4, including the outline of the area where dynamic changes have been made in the dynamic scenario.

In the dynamic scenario boxes of 0.6 m length, 0.4 m width and 0.25 m height are used to simulate a block storage in the center of the fixed moving path of the CTV. The height of the boxes is chosen to be lower than the mounting position of the SICK NAV350 sensor, such that the reference position is not affected by the dynamic changes. The block storage is build up, modified, and finally removed again in several steps. Using the recorded test data, both a localization with and without a map update are performed. In the following, results for an accuracy evaluation for the estimated position and orientation for a localization with and without map update are presented and compared. The accuracy estimation for both localization applications is based on the difference to a reference position obtained

Fig. 3 AGC used in test scenario

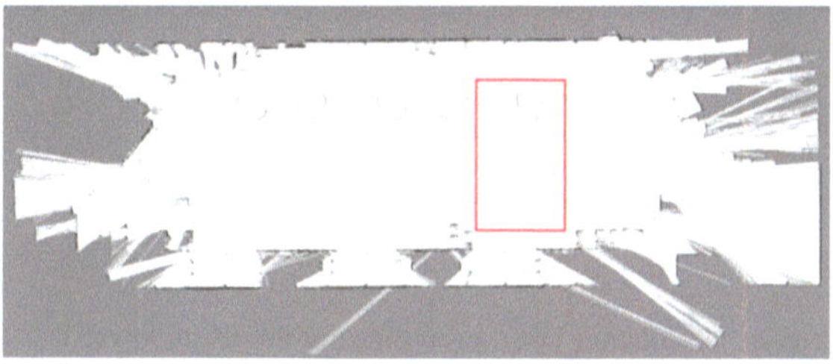

Fig. 4 Map of the test environment. Changes to the environment in the test scenario were made in the *red* highlighted area

by a reflector localization. The comparison of the position deviation for a localization is pictured in Fig. 5 and the equivalent deviation in orientation in Fig. 6. In both graphics vertical lines indicate points in time where changes in the environment have been made.

For a localization based on the static and non-updated map, the results show significant peaks for the position and orientation deviation. Hence, there are situations where the estimated pose for the CTV shows high uncertainties. These uncertainties occur when the CTV moves though the two created block storages and thus, they are obviously caused by the dynamic changes. For the complete test data the localization on a non-updated map results in a mean position error of 0.026 m and a standard deviation of 0.026 m. For a localization on a simultaneously updated map the mean position error is in a similar range at 0.022 m. A crucial difference however is realized for the standard deviation, which is almost halved to 0.014 m. Comparing a 3Ïƒ range for the position error emphasizes the improvement, as it comprises 99.87% of the expected errors in positions for tests results. Without any map update a range up to 0.10 m and with map update of 0.063 m can be expected. The decisive reason for this reduction is the elimination of peaks at positions where the localization based on a static map is highly influenced by dynamic objects (Fig. 5). In long-term applications in real-world logistic applications, the number and size of these peaks will increase for a localization without map update, resulting in problems for control systems and path planning. The proven ability of the proposed approach to eliminate these peaks allows for long-term reliable applications in industrial environments with many non-static objects. Similar findings result for the deviation in orientation for the test data. Without updating the map used for localization a mean of 0.515° and a standard deviation of 0.473°

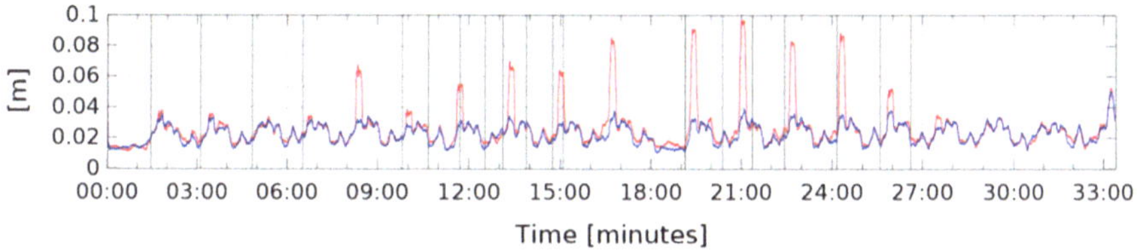

Fig. 5 Position error over time for localization on test data. With map update (*blue*) and without map update (*red*)

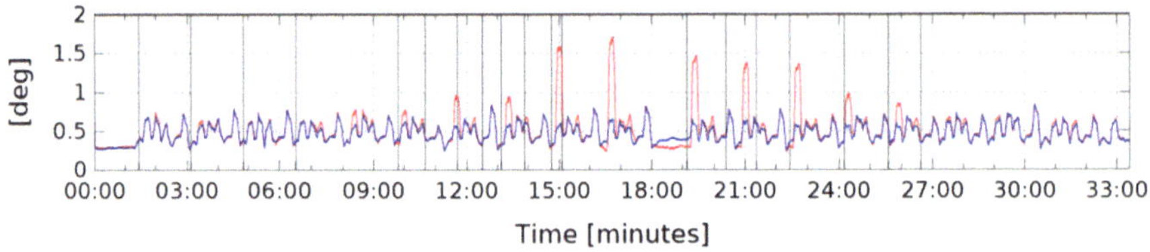

Fig. 6 Orientation error over time for localization on test data. With map update (*blue*) and without map update (*red*)

is achieved, resulting in a 3Ïƒ range of 1.93°. When updating the map using the proposed approach, these values can be reduced to a mean of 0.470°, a standard deviation of 0.305 and a resulting 3Ïƒ range of 1.39°. Again the plotted results for the orientation error in Fig. 6 underline the given quantitative results.

5 Conclusion

In this paper, we proposed an approach for localization of a mobile robot in dynamic environment, which updates the position of semi-dynamic objects in the map used in a following localization phase. The implemented algorithm provides a long-term reliable pose of a mobile robot in its environment even for over-time-changing environments. With the used sensor and the presented environment conditions it is shown that a long-term accuracy of well below 0.1 m is attainable. Moreover, the algorithm provides an updated map and an estimation of the accuracy of the determined pose with real-time capability. Nevertheless, the proposed approach makes basic assumptions on the dynamic characteristics of the environment. In order to guarantee for robust results, changes in the environment should be slow in the course of time, meaning that the observations of the sensors do not change to a significant extend within a short period of time. Second, we assumed that the movements of the robot allow for a regular observation of the complete environment for each single vehicle. In real world applications especially the second assumption might not always be given. Further development of the proposed approach that allow sharing information between multiple vehicles, offer potential for an additional optimization for real world applications. In a multi-robot scenario updates detected by a vehicle can be forwarded to other vehicles. A critical limitation for this development is the available bandwidth for transmitting information between the respective vehicles and a global control system. Potential approaches are sharing only actually updated pixels or sending only areas where changes have been detected, allowing other vehicles to adapt the update parameters for these area towards a faster update. By sharing information not all vehicles need to observe all parts of the environment regularly, but each part needs to be observed by at least one vehicle regularly.

References

Behnke J (2014) Logistische Regressionsanalyse: Eine Einfuehrung, Springer

Beinschob P, Reinke C, (2015) Graph SLAM based mapping for AGV localization in large-scale warehouses. In: IEEE International Conference on Intelligent Computer Communication and Processing (ICCP), Cluj-Napoca, pp. 245â€“248

Dellaert F et al (1999) Monte carlo localization for mobile robots. Robot Autom 2:1322–1328

Fernandez-Madrigal J A (2012) Simultaneous localization and mapping for mobile robots: introduction and methods, IGI global

Fox D (2003) Adapting the sample size in particle filters through KLD-sampling. Int J Robot Res 22(12):985–1003

Fujii A et al (2015) Detection of localization failure using logistic regression. In: Intelligent Robots and Systems (IROS), IEEE/RSJ International Conference on Robotics and Automation. Hamburg, pp. 4313â€“4318

Gutmann JS et al (1998) An experimental comparison of localization methods. Intell Robot Syst 2:736–743

Gustafsson F et al (2002) Particle filters for positioning, navigation and tracking. In: 11th European Conference on Signal Processing Toulouse, pp. 1â€“4

Kamagaew A et al (2011) Concept of cellular transport systems in facility logistics. In: 5th International Conference on Automation Robotics and Applications (ICARA). pp. 4045

Kirsch C et al (2002) Comparison of localization algorithms for AGVs in industrial environments

Kirsch C Roehrig C (2011) Global localization and position tracking of an automated guided vehicle. In: Proceedings of the 18th IFAC World Congress. Mailand

Kleiner A et al (2011) Armo: adaptive road map optimization for large robot teams. In: IEEE/RSJ International Conference on Intelligent Robots and Systems. San Francisco, pp. 3276â€“3282

Meyer-Delius D (2011) Probabilistic modeling of dynamic environments for mobile robots. PhD, Albert-Ludwigs-Universitt, Freiburg im Breisgau

Mongillo G, Deneve S (2008) Online learning with hidden Markov models. Neural Comput 20 (7):1706–1716

Pampel F C (2000) Logistic regression: A primer, vol 132, Sage Publications

Schulz D et al (2003) People tracking with anonymous and id-sensors using rao-blackwellised particle filters.In: IJCAI. pp. 921â€“928

Tipaldi GD et al (2013) Lifelong localization in changing environments. Int J Robot Res 32 (14):1662–1678

Valencia R et al (2014) Localization in highly dynamic environments using dual-timescale NDT-MCL. In: IEEE International Conference on Robotics and Automation, pp. 3956â€“396

Recognition of Lane Change Intentions Fusing Features of Driving Situation, Driver Behavior, and Vehicle Movement by Means of Neural Networks

Veit Leonhardt and Gerd Wanielik

Abstract The work presented aims at an early and reliable prediction of lane change maneuvers intended by the driver. For that purpose, an artificial neural network is proposed fusing features modeling the environmental situation that influences the formation of intentions, the gaze behavior of the driver preparing an intended maneuver and the movement of the vehicle. The sensor data required are provided by a multisensor setup comprising automotive radar and camera sensors. The whole prediction algorithm was put into practice as a real-time application and was integrated in a test vehicle. With this system, a naturalistic driving study was conducted on urban roads. The naturalistic driving data obtained were finally used for the parametrization of the algorithm by means of machine learning and for the evaluation of the prediction performance of the algorithm, respectively.

Keywords Lane change prediction · Intention recognition · Maneuver prediction · Sensor data fusion · Neural networks · Machine learning · Naturalistic driving data · Driver intention · Driver monitoring · Situation assessment

1 Introduction

With the rising number and complexity of vehicular systems aiming at assisting the driver and automating the task of driving, it becomes indispensable to automatically recognize and adapt the needs and intentions of the driver. This entails the need of algorithms that detect and assess driving situations as well as the behavior of the driver in real-time using vehicular sensor data. Accident statistics as for instance (Statistisches Bundesamt 2014) indicate that the need to assist the driver in par-

V. Leonhardt (✉) · G. Wanielik
Chemnitz University of Technology, Reichenhainer Strasse 70,
09126 Chemnitz, Germany
e-mail: veit.leonhardt@etit.tu-chemnitz.de

G. Wanielik
e-mail: gerd.wanielik@etit.tu-chemnitz.de

C. Zachäus et al. (eds.), *Advanced Microsystems for Automotive Applications 2017*,
Lecture Notes in Mobility, DOI 10.1007/978-3-319-66972-4_6

ticular applies to complex driving maneuvers, such as lane change maneuvers, which exhibit a significant higher accident risk. For that reason, previous scientific work already dealt with detecting and predicting lane change maneuvers by utilizing various types of features. These features differ in their reliability, their suitability for early prediction, and in the sensor data that have to be provided and processed. So (Kuge et al. 2000) and others worked on the detection of lane change maneuvers that had already started. That approach relied on parameters without exception that could be received directly from the vehicle's Controller Area Network (CAN), such as the steering force, the steering angle, and the steering angular velocity. But even if this could confirm the capability of those features to provide information regarding the presence of maneuvers, it has shown a lack in terms of a robust (Berndt and Dietmayer 2009) and early detection. Other studies as Oliver and Pentland (2000) and Leonhardt (2016) pointed out the suitability of environmental information about the geometry of the road and the surrounding traffic to improve the detection of upcoming lane changes. Parallel research, such as Henning (2010), Lethaus and Rataj (2007), Doshi and Trivedi (2009) and McCall et al. (2005), focused on the gaze behavior of the driver while preparing lane change maneuvers.

Previous work of the authors dealt with the robust as well as early prediction of upcoming lane change maneuvers based on the data fusion of characteristic features of various types. In doing so, the statistical analysis and comparison of individual features describing the driving situation, the driver's gaze behavior, and the movement of the vehicle presented in (Leonhardt and Wanielik 2017) could verify the complementary strengths of those features in connection with predicting different types of lane change maneuvers and with regard to the time of prediction. In (Bengler et al. 2017) and (Leonhardt et al. 2016), the authors could prove that combining features of the driving situation and the gaze behavior of the driver by means of a Bayesian network can significantly improve the overall performance of the lane change prediction with respect to an early as well as reliable prediction. However, the mainly rule-based parametrization of the network left aside potential for further optimization as well as not involving the vehicle's movement.

Thus, the present work aims at improving the feature fusion of the lane change prediction using machine learning in connection with naturalistic driving data. Accordingly, this paper starts with the introduction of the feature set used for the prediction in Sect. 2. Details that relate to the structure of the prediction algorithm, the sensor setup used for data acquisition, and the computation of the features are briefly described in Sect. 3, as Sect. 4 is about the naturalistic driving study and its results. Section 5 deals with the design and parametrization of the artificial neural network which is applied. Section 6 covers the evaluation of the algorithm, followed by Sect. 7 concluding with findings and future work.

2 Features Indicating Upcoming Lane Changes

Predicting lane change maneuvers intended by the driver relies on detecting indications of one of the three consecutive phases: The formation of the intention to change the lane, the preparation of the lane change, and the maneuver action. The first one is a result of the driver's perception and assessment of the nearby environment and the surrounding traffic situation. If another lane is available and is considered to be more attractive for driving, the driver will intend to change to that lane. The following preparation of the driver to carry out a safe lane change maneuver is accompanied by an intensified gaze behavior that is proven to be characteristic by research as done by Lethaus and Rataj (2007) and Doshi and Trivedi (2009). Finally, the driver starts to perform the lane change which results in changes of the vehicle's status and movement parameters. All these phases cause measurable parameters to change in a specific manner. So feature values can be derived indicting upcoming maneuvers, even if the intention itself cannot be observed directly.

The features used for the intention recognition algorithm presented are spread over three subsets of features and refer to the previous work of the authors. It can be found in Leonhardt and Wanielik (2017) and deals with the evaluation of features of all of these phases with respect to their ability to predict lane change intentions early and in different situations.

The first subset of features proposed aims at assessing the driving or traffic situation based on the estimation of the lane's occupancy. It is assumed that the driver tends to avoid uncomfortable or dangerous situations. So obstacles on the lane the driver is driving can be considered as a motivating factor to change the lane. By contrast, the absence of another lane or heavy traffic on it can impede the formation of such an intention (Schroven and Giebel 2008). The degree of occupancy is modeled by means of the distance, relative velocity, and adapting Deceleration to Safety Time (DST) in relation to the closest vehicle on the lane the driver is driving and to the closest vehicles on the adjacent lanes in front of and behind the driver, respectively. In this context, the adapting DST is seen as a measure of discomfort by harsh deceleration. It is defined as the degree of deceleration that would be needed to meet a minimum safety distance of at least 2 seconds of driving in relation to a moving object. For a detailed derivation, please refer to Bengler et al. (2017). Moreover, the lane accessibility is utilized as a feature to quantify the possibility of a legal lane change. A maneuver is considered being legal in case the corresponding adjacent lane exists and is not restricted by solid lane markings.

The second feature subset consists of six features modeling the gaze behavior of the driver by means of the head's three-dimensional position and orientation in the directions of the axis x, y and z.

The third subset describes the status and current movement of the vehicle. It consists of the lateral distance to the closest lane marking and of parameters that can be received directly from the CAN bus of the vehicle: the level of the actuation of the gas pedal, the break pressure, the velocity, and the yaw rate of the vehicle. With the exception of the first one and in contrast to the features mentioned before, their usage generally does not entail the need of additional sensor systems and more complex feature calculation. Another feature could be the indicator signal indicating lane change intentions by definition. However, it was decided not to apply it in order to evaluate the algorithm's performance regardless its frequency of use.

3 Implementation and Sensor Data

The algorithm for lane change intention recognition being proposed was realized as a real-time application running under real conditions in a test vehicle. It comprises software modules for sensor data acquisition, features computation and a neural network processing the features to conclude on upcoming lane change maneuvers as depicted in Fig. 1.

The test vehicle is equipped with a total of six automotive radar sensors covering 360° of its environment. The radar data serve to detect and track moving objects around and form the basis for the features assessing the driving situation. In doing so, the detected objects are grouped into zones that are located around the vehicle. The zones classify the objects to be in front of the vehicle or behind it and thereby distinguish between the lane of the vehicle and the closest adjacent lanes to the left and right, respectively. For each zone, a reference object is selected by means of distance. These reference objects serve to calculate the features assessing the occupancy of the corresponding areas. To allocate the objects tracked to the zones, knowledge about the curvature of the lanes and the relative position of the vehicle within its lane is needed. That knowledge is provided by a commercial camera-based system called Mobileye which is supplemented by a detection of the type of the lane markings to determine whether or not a legal lane change would be possible. As stated before, the status and movement parameters of the ego vehicle are provided by the CAN bus. In doing so, the velocity and yaw rate are used to track the motion of the vehicle as well as to support the object tracking.

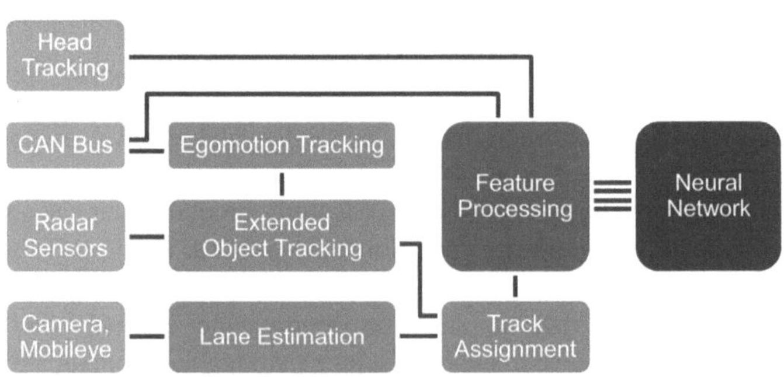

Fig. 1 Structure of the entire algorithm predicting upcoming lane changes

Furthermore, the test vehicle has several camera sensors. One of them is facing the driver and supplies video data for the head tracking that is based on a commercial system called Seeing Machines faceAPI. For a more detailed description of the technical setup and the preprocessing of the sensor data, please refer to (Leonhardt et al. 2016). Finally, all features are calculated and passed on to the neural network implemented to classify whether or not there is an imminent lane change maneuver.

4 Naturalistic Driving Study

For the purpose of optimal parametrization with the test vehicle, a naturalistic driving study was conducted. It took place under real traffic conditions without the presence of any student manager and any hints referring to the study's background or focus. The 60 participants were only instructed to follow a 40 km long predefined route of mostly two-lane roads in the urban area of Chemnitz. A more detailed view on the study can be found in Leonhardt and Wanielik (2017).

All sensor data were recorded and annotated with regard to the time and the direction of the lane change maneuver and with regard to the traffic situation that led to the maneuver. Figure 2 gives an overview on the types of maneuvers that occurred and on their absolute frequency. It can be seen that most of the 1,869 lane change events are connected with situations in which a [Slow Vehicle] driving in front is overtaken by means of a lane change to the left. This also includes overtaking parking vehicles and stationary vehicles at traffic lights. By contrast, lane changes triggered by permanent obstacles as construction areas are categorized as [Static Obstacle]. Finishing overtaking maneuvers with a lane change to the original lane is labeled as [Return Lane]. Where the road is extended by another lane, changing to that lane is designated as [Added Lane]. Furthermore, [Enter Lane] denotes lane changes to enter a road, e.g., from a feeder road. In the urban scenario of the study, all added lanes were turn lanes and all feeder roads approached from the right. Lane changes triggered by other vehicles entering the road are classified as [Merging Vehicle]. All remaining change maneuvers without any identifiable reason are tagged with the label [Unknown].

5 Neural Network for Feature Classification

As an extension of the authors' previous work (see Bengler et al. 2017) focusing on the prediction of lane changes induced by vehicles driving ahead slowly, the prediction algorithm is generalized to all types of lane change maneuvers. Since this requires a more balanced and generalizable adjustment of the algorithm, it was decided to apply machine learning in combination with the naturalistic driving data obtained by the driving study. For this purpose, artificial neural networks were utilized. They are a convenient approach to solve such problems of classification

Lane Change Type	Pictogram	Number of Lane Changes	
Slow Vehicle	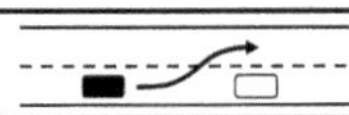	$n_{LEFT} = 507$	
Return Lane	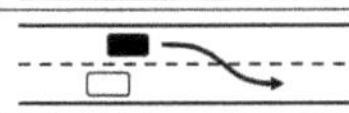		$n_{RIGHT} = 520$
Added Lane	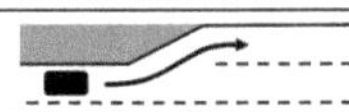	$n_{LEFT} = 242$	$n_{RIGHT} = 242$
Enter Lane		$n_{LEFT} = 239$	
Merging Vehicle	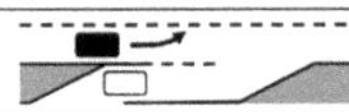	$n_{LEFT} = 33$	
Static Obstacle	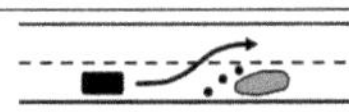	$n_{LEFT} = 19$	
Unknown	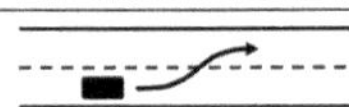	$n_{LEFT} = 61$	$n_{RIGHT} = 6$
		$n_{TOTAL} = 1869$	

Fig. 2 Lane change maneuvers observed in connection with the driving study

fusing multiple input parameters and they can rely on sophisticated implementations and tools for machine learning.

5.1 Artificial Neural Networks

By analogy with the behavior of biological brains, artificial neural networks model the statistical relation of input variables and output variables by a collection of neurons and neuron connections. The neurons of such networks are typically grouped in layers as can be seen in Fig. 3. Input signals enter the network via the input layer, pass a variable number of possibly different types of hidden layers, and

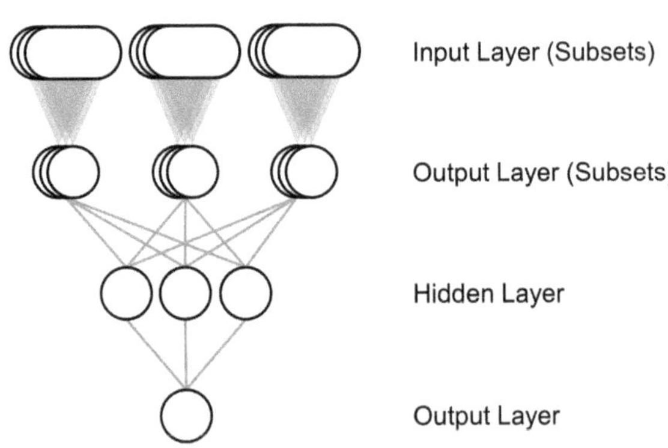

Fig. 3 Structure of the artificial neural network proposed for lane change prediction

finally reach the output layer. While passing the layers, the input signals are repeatedly modified and combined to finally obtain the aspired output value.

In the case of dense or fully connected layers, each neuron of a layer is individually connected to each neuron of its preceding and following layer, respectively. Differences in the dependence of the neurons are modeled by means of individual weighting factors describing the extent of connection. In accordance with the sign of this factor, enhancing dependencies can be modeled as well as inhibiting ones. The state of any neuron is represented by its activation. Provided that a neuron is not an input neuron, this activation value is obtained by summing up its bias value and the activation values of all neurons of the preceding layer weighted by their extents of connection. Subsequently, the weighted sum is modified applying an activation function to limit the activation value and to manage the propagation behavior. The parametrization is usually done by supervised learning from data rather than by explicit parametrization. To sum up, a neural network's performance strongly depends on the set of features used not only, but also on the suitable design of the network, the process of training, and the data used.

5.2 Network Design

The approach provides for two neural networks in parallel to predict lane changes to the left and right separately. The design of these networks proposed to fuse the features presented and to deduce lane change intentions is depicted in Fig. 3.

It comprises three network sections to fuse the features of each of the three subsets individually and one section to combine the results of the other sections. The latter includes one hidden layer of three neurons, the others do not contain any additional layers. Deviating network designs with a varying number of layers and neurons per layer were evaluated as well. Thereby, it has become evident that they do not necessarily show a significant better prediction performance for the present classification problem, despite their higher complexity. This kind of subdivision was chosen to evaluate the prediction performance using features based on the environment, the gaze behavior of the driver and the status of the vehicle separately. All layers within the sections are fully connected. The activation function used is sigmoid.

The evaluation of different network designs also pointed out that the prediction performance benefits from information referring to the progression of the values of the features over time. However, including the features' complete history over a timespan of a few seconds turned out to make the network disproportionately complex for the problem to be solved so that the network tends to overfit while training. Overfitting describes a neural network's behavior to model the data used for training so exactly that it loses its ability to generalize. It performs well in connection with data known but poor with the data unknown. Nevertheless, to take into account the features' history in a lean as well as effective manner it is proposed to model the past few, in this case 10 seconds of the features by means of their minimum and maximum value that occurred over that period of time. Consequently, together with the current value, each feature is modeled by a total of three input neurons.

5.3 *Network Parameterization*

For the sake of parameterizing and evaluating the neural networks, the data obtained by the driving study were divided into two data sets. The first data set contains about 70% of the data and was used to train the networks. The remaining 30% of the data were used for evaluation only. While dividing the data it was ensured that the types of drivers, their number as well as the number of the maneuvers occurred were spread across the data sets in the same proportion. Furthermore, the data of none of the drivers were used in both data sets in order to evaluate the algorithm's performance concerning those drivers the algorithm was not trained for.

The parameterization was done by applying backpropagation on the training data with 15 Hz utilizing the open source neural network library Keras with the open source library TensorFlow as a backend. The output value that was trained complies with the assignment of the input data to one out of two classes indicating whether or not the data are related to a situation of an imminent lane change maneuver. The delimitation of the two classes is made in accordance with the distance in time to the next lane change maneuver. Following (Lethaus et al. 2011) all measurements are assumed to be related to an imminent lane change situation that occur over a period of 10 s prior to the moment the vehicle's longitudinal axis passes the lane markings of its lane. Moreover, the first 3 s of the data after passing the lane markings are excluded from the training and evaluation being part of lane change maneuvers that are already in progress. The comparison of the expected and the predicted classification during the process of training was done by means of a custom loss function. It is designed to balance the prediction performance of both classes to prevent the training from being dominated by the much more frequent situations without any imminent lane change. Furthermore, it allows to train a much more natural transition between the classes.

6 Experimental Results

To evaluate the ability of the entire algorithm to predict upcoming lane changes, the prediction performance was analyzed applying the receiver operating characteristic (ROC) method on the evaluation data set not used before. The ROC contrasts the ability of the algorithm to detect situations of upcoming lane changes reliably with its probability to give rise to false alarms in different situations. The value under test was the final output of the algorithm concluding with the classification done by the neural network. The variable parameter used to adjust the classification performance was the discrimination threshold that is applied on the output value of the network to discriminate the two classifications being possible. Furthermore, in order to assess the algorithm with regard to an early prediction, the prediction performance using different amounts of data was evaluated. For that purpose, in each case, all measurement data were excluded from being used for the training and the evaluation that occurred within a specified period of time before the vehicle passes the lane markings. In doing so, the algorithm was forced to predict the maneuvers not later than that time. The earliest point in time maneuvers could be predicted was by definition 10 s before passing the markings. Each prediction before was considered as false alarm.

Sections (a)–(e) of Fig. 4 show the resulting receiver operating characteristics of the network parameterized to predict lane changes (LC) to the left for the various times of prediction. As might be expected, the most reliable prediction results can be achieved by using all feature subsets and taking into account all information until the maneuver happens. So according to section (a), the algorithm performs with an area under curve (AUC) of 0.964. For a false alarm rate of 0.020, for example, about 98.3% of the lane change maneuvers to the left can be predicted before the vehicle passes the lane markings. Moreover, 98.3 and 94.2% of the maneuvers can already be predicted 2 and 4 s earlier at the latest with false alarm rates of 0.038 and 0.063, respectively. 8 s before changing the lane a prediction rate of 71.6% with a false alarm rate of 0.089 can still be achieved. A comparison of the performance of the individual feature subsets of driving situation, maneuver preparation, and vehicle status shows that each subset is individually suitable for predicting upcoming lane changes. Thereby, the features modeling the status and movement of the vehicle show the most distinctive performance. They may also be less affected by measuring errors than the others. With regard to the remaining two feature sets, in section (f) of the figure, it can be seen that the feature set representing the driving situation is more suited to detect the early phase of situations of changing the lane. On the other hand, information regarding the movement of the head of the driver provide better results at the later phase. This is consistent with the findings of Leonhardt and Wanielik (2017) evaluating the features statistically. But in addition, it appears that fusing the features of all three sets outperforms all individual sets in any case. In comparison with using the subsets individually, combining them turns out to be able to reduce the false alarm rate by up to 77%.

7 Conclusion and Future Work

A complete approach was presented to predict lane change maneuvers intended by drivers early as well as reliably. The approach relies on three sets of features derived from vehicular sensor data describing the driving situation before lane changes, the driver's behavior preparing a maneuver and the vehicle's status, respectively. To achieve a robust prediction performance, it was proposed to fuse the features by

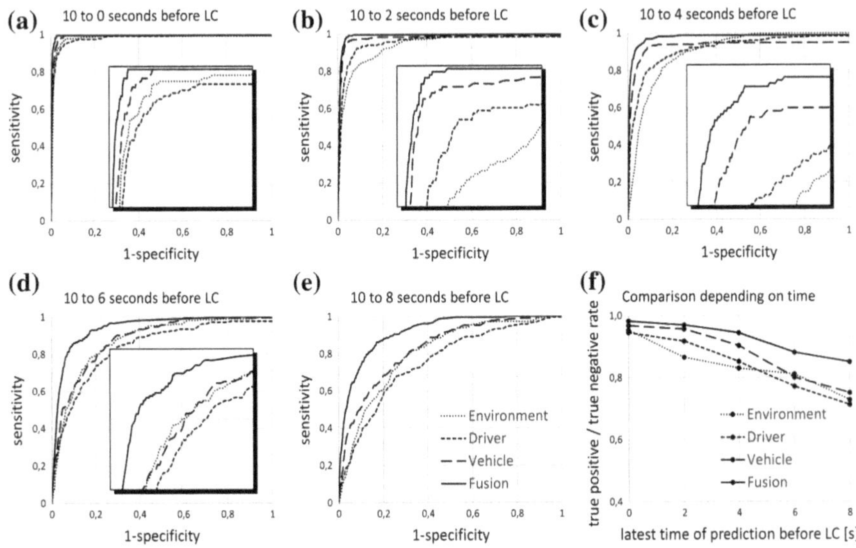

Fig. 4 Prediction performance depending on time and feature set (evaluation data)

means of an artificial neural network optimized by machine learning. The network comprises a compact way to involve the temporal course of the features which improves the performance of the prediction and at the same time limits the number of neurons to a minimum. This inhibits early overfitting, the lane change situations the network was trained with and preserves the capability for generalization. The resulting algorithm was put into practice as a vehicular application doing the sensor data acquisition, feature preprocessing, and the maneuver prediction in real time. The parametrization of the network as well as the evaluation were done on the basis of naturalistic driving data obtained by a driving study conducted. The realization of the algorithm as well as the evaluation of its performance have proven that the approach is suited to predict lane change maneuvers in an early and reliable manner. Moreover, it could be shown that the prediction significantly benefits from using and fusing features of all three categories.

In comparison with prior results as in Leonhardt et al. (2016) and Bengler et al. (2017), the lane change prediction, on the one hand, was extended to predict lane change maneuvers in general situations and at the same time, it was improved in terms of performance. Primarily, this was achieved by involving the status and movement of the vehicle as proposed in Leonhardt and Wanielik (2017) and by applying machine learning in connection with naturalistic driving data. However, it can be noticed that a few situations still are difficult to detect. This could be an indication for the need of additional training data covering more driving situations and individual driver characteristics. So, as might be expected, more driving data would probably additionally improve the prediction performance. Moreover, further work could aim at determining and leaving out features of lesser or redundant influence.

Acknowledgements This work bases on results of the research project UR:BAN Internet Presence (2017). With its 30 partners, it aimed at developing user-oriented assistance systems and network management in urban space. It was supported by the Federal Ministry of Economics and Technology on the basis of a decision by the German Bundestag.

References

Bengler K, Drüke J, Hoffmann S, Manstetten D, Neukum A (2017) UR:BAN human factors in traffic—approaches for safe, efficient and stress-free urban traffic. In: Leonhardt V, Pech T, Wanielik G (eds) Chapter fusion of driver behaviour analysis and situation assessment for probabilistic driving manoeuvre prediction. Springer Vieweg, Heidelberg. ISBN 978-3-658-15417-2

Berndt H, Dietmayer K (2009) Driver intention inference with vehicle onboard sensors. In: Proceedings of the IEEE international conference on vehicular electronics and safety (ICVES), pp. 102–107

Doshi A, Trivedi MM (2009) On the roles of eye gaze and head dynamics in predicting driver's intent to change lanes. IEEE Trans Intell Transp Syst 10(3):453–462

Henning M J (2010) Preparation for lane change manoeuvres: behavioural indicators and underlying cognitive processes, Ph.D. dissertation, Chemnitz University of Technology, Chemnitz

Kuge N, Yamamura T, Shimoyama O, Liu A (2000) A driver behavior recognition method based on a driver model framework, SAE Technical Paper 2000-01-0349

Leonhardt V, Wanielik G (2017) Feature evaluation for lane change prediction based on driving situation and driver behavior. In: Proceedings of the IEEE international conference on information fusion

Leonhardt V, Pech T, Wanielik G (2016) Data fusion and assessment for maneuver prediction including driving situation and driver behavior. In: Proceedings of the IEEE international conference on information fusion, pp. 1702–1708

Lethaus F, Rataj J (2007) Do eye movements reflect driving manoeuvres? Intell Transport Syst (IET) 1(3):199–204

Lethaus F, Baumann MRK, Köster F, Lemmer K (2011) Using pattern recognition to predict driver intention. Springer, Heidelberg

McCall J C, Trivedi M M, Wipf D, Rao B (2005) Lane change intent analysis using robust operators and sparse Bayesian learning. In: IEEE computer society conference on computer vision and pattern recognition (CVPR)—Workshops, pp. 59–59

Oliver N, Pentland A P (2000) Graphical models for driver behavior recognition in a SmartCar. In: Proceedings of the IEEE intelligent vehicles symposium (IV), pp. 7–12

UR: BAN Internet Presence (2007) "http://urban-online.org/de/urban.html

Schroven F, Giebel T (2008) Fahrerintentionserkennung für Fahrerassistenzsysteme. In: Proceedings of 24. VDI/VW-Gemeinschaftstagung—Integrierte Sicherheit und Fahrerassistenzsysteme, Wolfsburg Bd. VDI-Berichte, VDI Verlag, Düsseldorf

Statistisches Bundesamt (2016) Verkehr—Verkehrsunfälle 2014, Fachserie 8, Reihe 7

Applications of Road Edge Information for Advanced Driver Assistance Systems and Autonomous Driving

Toshiharu Sugawara, Heiko Altmannshofer and Shinji Kakegawa

Abstract As road edge information gives various benefits to ADAS (Advanced Driver Assistance Systems) and AD (Autonomous Driving) applications, we have developed a road edge detection algorithm with stereo camera. In this paper, two ADAS/AD applications using road edge information, validated by simulation and experiment, are introduced to show benefits of the method. An integrated lateral assist system including LDP (Load Departure Prevention system) and RDP (Road Departure Prevention system), which assumes lane support system in Euro NCAP, is proposed as an ADAS application. Also, "obstacle avoidance through road shoulder space" is introduced as an application for AD.

Keywords Stereo camera · ADAS · AD · Euro NCAP · LDP · RDP · Obstacle avoidance

1 Introduction

For reducing traffic accidents and driver's workload, ADAS and AD are being developed all over the world. LDP and LKAS (Lane Keeping Assist System) are some of the most popular applications of ADAS/AD. These applications inform the driver about departure with haptic, acoustic or visual information and/or control actuators with position information of lane markings detection by cameras or

T. Sugawara (✉) · H. Altmannshofer
Automotive and Industry Laboratory, Hitachi Europe GmbH, Tecnopark IV Lohstrasse 28, Schwaig-Oberding, Germany
e-mail: Toshiharu.Sugawara@hitachi-eu.com

H. Altmannshofer
e-mail: Heiko.Altmannshofer@Hitachi-eu.com

S. Kakegawa
Centre of Technology Innovation, Smart System Research Department, Hitachi, Ltd., Ohmika-Cho 7-1-1, Hitachi-Shi, Ibaraki, Japan
e-mail: shinji.kakegawa.od@hitachi.com

C. Zachäus et al. (eds.), *Advanced Microsystems for Automotive Applications 2017*,
Lecture Notes in Mobility, DOI 10.1007/978-3-319-66972-4_7

LIDAR. However, lane markings are sometimes not painted, especially in country roads, or may not be clearly visible due to weather or road condition. In these cases, the above applications might not work properly since lane makings are not detected. If road edges are detected, the applicable scene for LKAS and LDP will be expanded since the system can keep the function without lane markings.

We have developed a road edge detection algorithm with stereo camera in order to expand applicable scenarios of the mentioned systems (Kakegawa and Sugawara 2016). In this paper, two ADAS/AD applications using road edge information, validated by simulation and experiment, are introduced to show benefits of road edge information.

The integrated lateral assist system including LDP and RDP, which assumes lane support systems in Euro NCAP (Williams 2017), is introduced as an ADAS application. There are a lot of researches and developments related to LDP and RDP. Nissan was the first to provide LDP to the market (Yamamura 2008). RDP system concept and maximum vehicle steering capability with ESC (Electronic Stability Control) was mentioned in (Lattke et al. 2015). However, in the previous research, how to integrate LDP and RDP and how to coordinate EPS (Electric Power Steering) and ESC were not clearly mentioned. Therefore, the integrated lateral assist system including LDP and RDP with EPS and ESC is proposed taking into account characteristic of actuators and application in this paper. Experimental results show the validity of the proposed system.

Several path planning algorithms for AD and applications are already introduced (Kuwata et al. 2009; McNaughton et al. 2011; Dolgov et al. 2008; Montemerlo and Becker 2008; Howard et al. 2008). However, in the previous research, the benefits of road edge information for AD are not clearly mentioned. Therefore, "obstacle avoidance through road shoulder space" is introduced as an application for AD. The experimental result of autonomous obstacle avoidance shows that the system can judge road shoulder space as safe area with road edge information.

2 Road Edge Detection

2.1 Target Road Edge

The various kinds of road edges are plotted schematically in Fig. 1. The target road edges in this paper are the road edges with positive height irrespective to the kinds of road edges. The current target range of height is the height over 10 cm. This is based on the knowledge that the one of the most important road edges with small height is the curbstone, and the typical height of the curbstone is over 10 cm.

Stereo camera is used for road edge detection, since stereo camera has a good potential as a sensor used for road edge detection. Stereo camera can output a disparity map in addition to the captured images. The disparity map contains distance information to objects with high space resolution. Therefore, we have

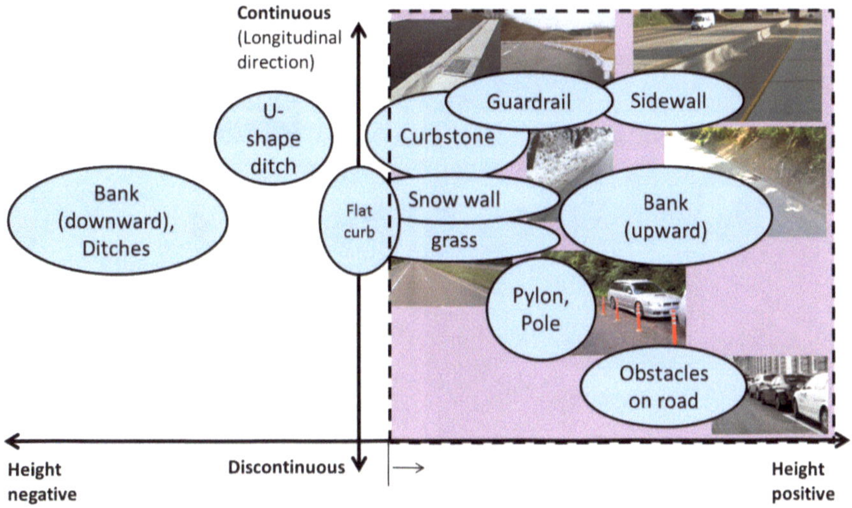

Fig. 1 Target road edge

developed the road edge detection algorithm with stereo camera (Kakegawa and Sugawara 2016).

2.2 *Road Edge Detection Result*

It was confirmed that the various kinds of road edges were stably detected, including sidewall, guardrail, curbstone, grass, and bank. The periodic road edge or discontinuous road edge such as pole and side edge of vehicle were also well detected. The examples of the detection result are shown in Fig. 2. In the figure, the overlaid line with rectangles represents the detected road edge, and the hatched area represents the computed free space. It is found that the free space is also well recognized since the lateral boundary of the free space is correctly delimited at the location of the road edge including the edges with low height such as curbstone.

3 Application for Advanced Driver Assistance Systems

3.1 *Euro NCAP*

Euro NCAP is considering to expand the test protocol of lane support system in 2018 (Williams 2017). The expanded test protocol includes Lane Departure Warning (LDW), Lane Keep Assist (LKA), and Emergency Lane Keep (ELK). As

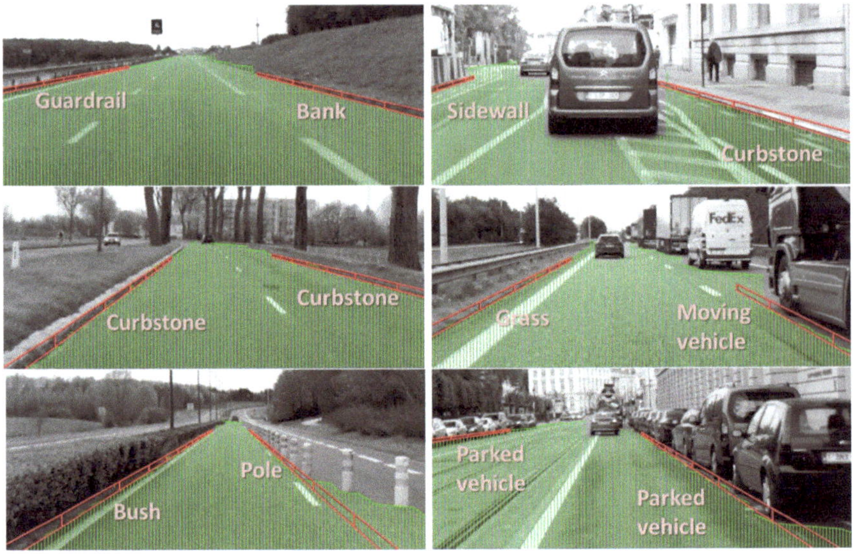

Fig. 2 Detection result of road edge

shown in Fig. 3, the test protocol of ELK includes (1) the vehicle departs to road edge with/without lane markings, (2) the vehicle departs to the lane where there is the oncoming vehicle, and (3) the vehicle departs to the lane where there is the overtaking vehicle. In test protocol of LKA, the vehicle depart to road edge or solid/dashed line is assumed as shown in Fig. 4. In test protocol of LDW, the vehicle depart to solid/dashed line is considered as shown in Fig. 5.

3.2 Integrated Lateral Assist System

3.2.1 Overview of Virtual Lane Guide

We have proposed the integrated lateral assist system called "VLG (Virtual lane guide)" (Nakada et al. 2011; Yokoyama and Saito 2009). VLG concept is shown in Fig. 6. Based on lane marking, road edge, and other obstacle information, VLG applied moment to avoid crash with the oncoming/overtaking vehicle and unintentional lane/road departure depending on the risk. VLG calculates two types of virtual lane. One is the low-risk virtual lane calculated along lane makings. When the vehicle is approaching to the low-risk virtual lane, small moment is applied with beeping sound in order to initiate driver's counter steering maneuver. Another is the high-risk virtual lane calculated along road edge and around the oncoming/overtaking vehicle. When the vehicle is approaching to the high-risk

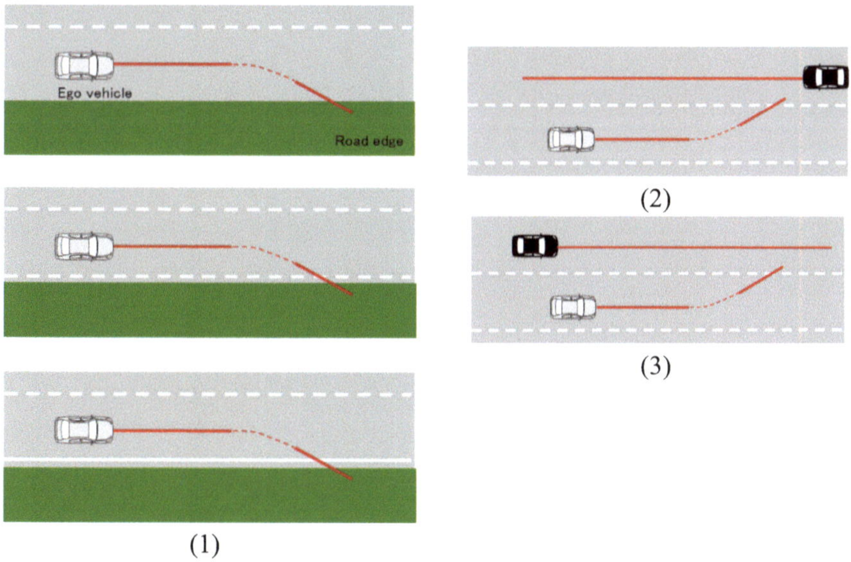

Fig. 3 Emergency Lane Keep: (*1*) departure to road edge, (*2*) departure to the lane where there is the oncoming vehicle, (*3*) departure to the lane where there is the overtaking vehicle

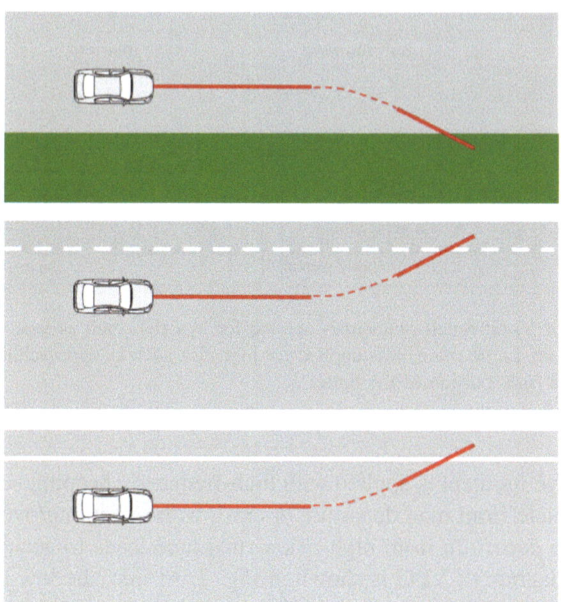

Fig. 4 Lane keep assist: The *top figure* is additional road edge scenario. The *middle* and *bottom figure* is scenario for lane departure

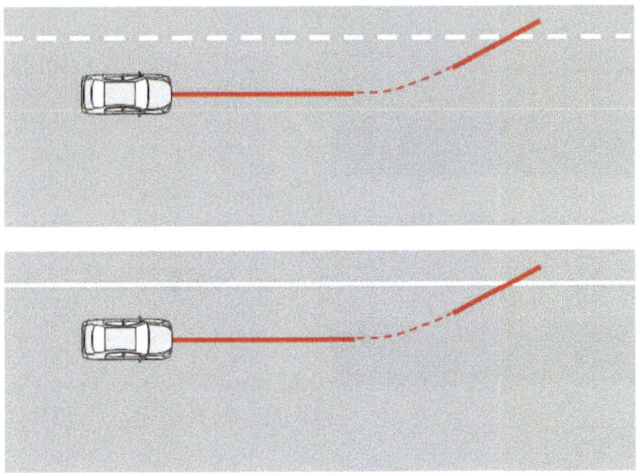

Fig. 5 Lane departure warning

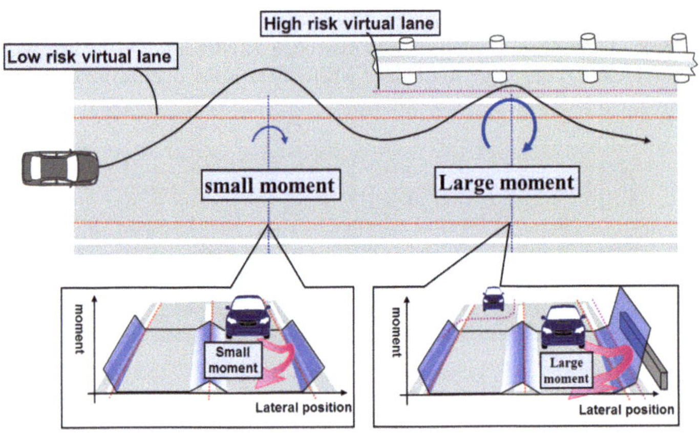

Fig. 6 Concept of VLG: Small moment is applied for low risk such as lane departure like *left* portion of the figure. Large moment is applied for high risk such as approaching to road edge or other vehicles like *right* portion of the figure

virtual lane, large moment is applied with high-frequency beeping sound in order to prevent the vehicle from road departure or crash with oncoming/overtaking vehicle surely. Because departure from high-risk virtual lane leads to accidents directly.

The block diagram of VLG is shown in Fig. 7. At first, the low-risk virtual lane is calculated based on lane markings. The high-risk virtual lane is calculated by obstacle and road edge information. In parallel, the preview point is calculated to estimate the vehicle's future position based on steering angle and velocity. By

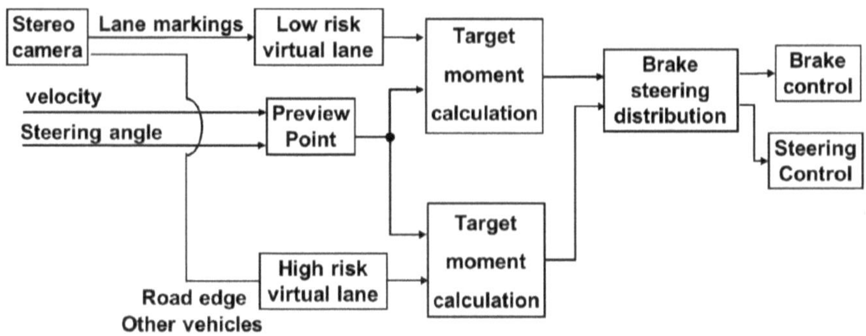

Fig. 7 Block diagram of VLG

Table 1 Target of VLG

Function	Departure Prevention from low-risk virtual lane	Departure Prevention from high-risk virtual lane
Use case	Lane departure	Road departure Lane departure toward obstacle
Frequency	High	Low
Risk	Low	High
Target of control	(1) initiate driver's correction maneuver within not intrusive intervention (2) can assist many times	(1) Prevent the vehicle from departure from the high-risk virtual lane surely.

comparing the preview point with each virtual lane, each target moment is calculated. At first, the target moment of low- or high-risk virtual lane which is higher than another one is selected in the brake steering distribution block. Then, the final target moment is distributed to EPS and ESC using a coordination method mentioned in Sect. 3.2.3.

3.2.2 Target of VLG

Next, target of VLG is described as shown in Table 1. The departure from the low-level virtual lane is high frequency in daily life, however, the lane departure is not directly lead to accidents. Thus, it is decided that "target of departure prevention from low risk virtual lane" is: (1) initiate driver's correction maneuver within not intrusive intervention. (2) can assist many times. In other words, intervention for lane departure prevention should have high durability. On the other hand, departure from high-risk virtual lane is very low frequency. However, the departure directly leads to accidents. Therefore, it is decided that "target of departure prevention from the high risk virtual lane" is to prevent the vehicle from departure from the virtual lane surely.

3.2.3 Coordination of EPS and ESC

The characteristic of actuators is considered to decide how to coordinate ESC and EPS for departure prevention system as shown in Fig. 8. Moment with ESC to steer vehicle makes deceleration feeling, wearing brake pad, and increases fuel consumption due to recovery of speed. However, ESC can steer vehicle without interference to the driver's steering control. On the other hand, EPS can steer vehicle without deceleration feeling, increasing fuel consumption and wearing components, However, steering assist affects driver's steering wheel control. The steering assist torque should be limited to avoid intrusive intervention.

Based on the target of VLG and the characteristic of actuators, the coordination method of ESC and EPS are proposed as shown in Fig. 9. In case of low risk such as lane departure and start of road departure, moment is mainly distributed to EPS to correct driver's unintentional departure without much deceleration feeling and fuel consumption. In case of high risk such as approaching road edge, target moment is mainly distributed to ESC to avoid crash surely without much interference to driver. The distribution ratio and the maximum steering assist torque can be changed.

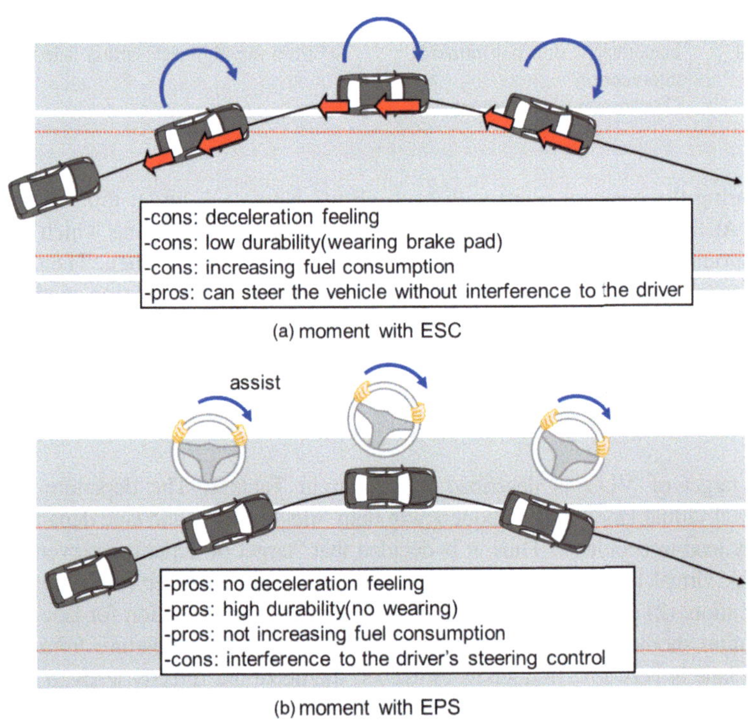

Fig. 8 Characteristic of actuators

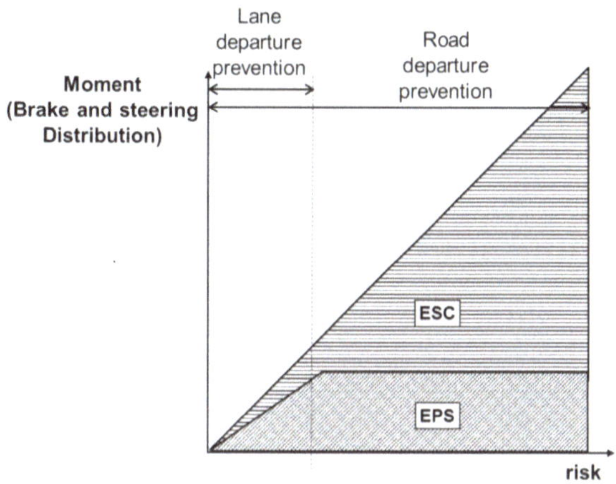

Fig. 9 Brake steering distribution

3.3 Experimental Result

Experiments were conducted with our test vehicles in which Hitachi's stereo camera, EPS and ESC are installed. At first, lane departure prevention was tested under the condition shown in Fig. 10. The testing result is shown in Fig. 11. The target moment was distributed to EPS mainly based on the above-mentioned coordination method. The vehicle departed from lane for a moment, however the VLG could keep the vehicle inside the ego lane eventually without much braking (deceleration feeling and fuel consumption).

Next is a testing result of road departure prevention conducted under the condition described in Fig. 12. When the risk was high (vehicle approached to the road edge) from 0.6 [s] to 1.3 [s], the target moment was mainly distributed to ESC. It was confirmed that VLG can avoid road departure surely with the moment by ESC and limited steering assist (without much interference to the driver) (Fig. 13).

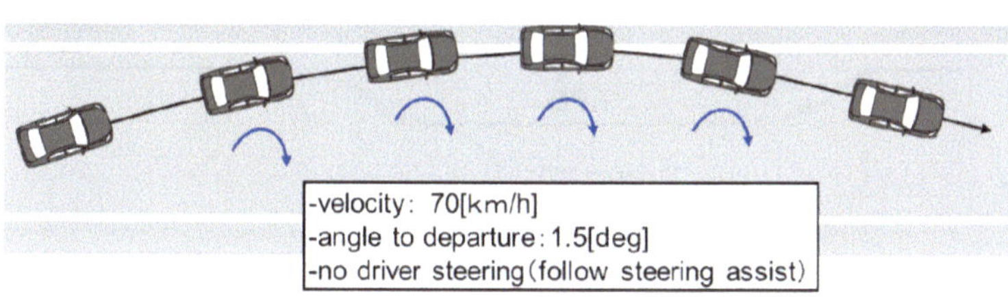

Fig. 10 Test condition for lane departure prevention

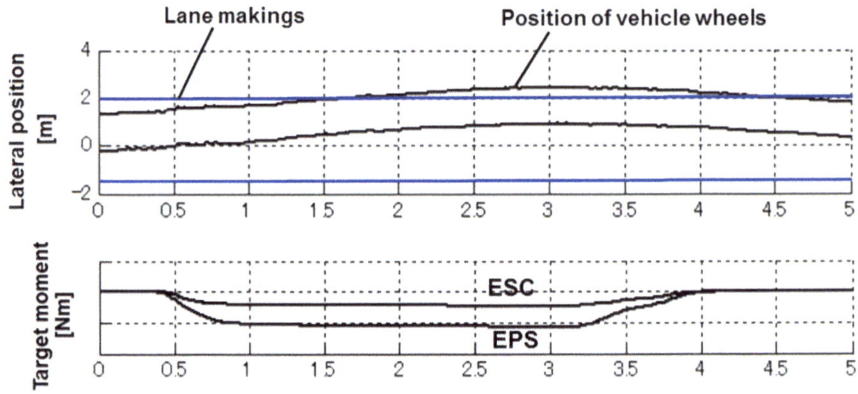

Fig. 11 Experimental result for lane departure prevention

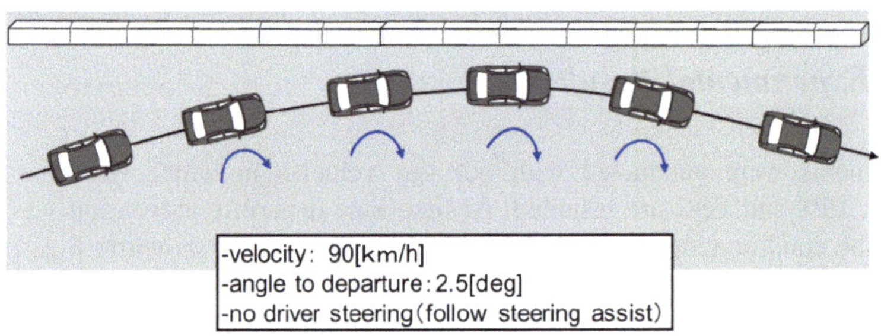

Fig. 12 Test condition for road departure prevention

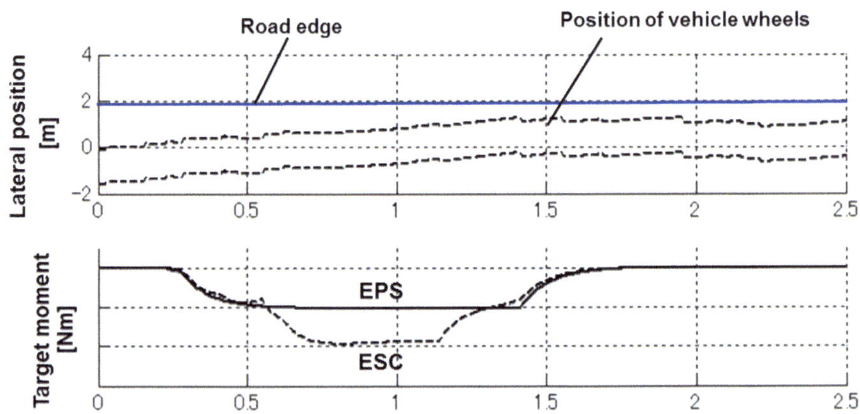

Fig. 13 Experimental result for road departure prevention

4 Application for Autonomous Driving

4.1 Path Planning Algorithm

In this section, autonomous obstacle avoidance is introduced to show the benefit of road edge information for AD. At first, the block diagram of path planning algorithm is shown in Fig. 14. It consists of three blocks of path planner, path selector, and path follower. The path planner calculates candidate paths by sampling in state space (Montemerlo and Becker 2008; Howard et al. 2008). The path selector selects an optimal path from the candidate paths taking into account navigation route, riding comfort, traffic rules, and collision risks to obstacle and road edge. The path follower calculates steering assist torque based on the optimal path.

4.1.1 Path Planner

The path planner generates several paths with lateral offset to center of lane by state space sampling. The ego vehicle can change path to left or right with respect to the center of lane by changing path (lateral offset). By intelligently changing the lateral offset, ego vehicle can achieve several use cases such as auto merging, auto overtaking, auto branch, obstacle avoidance, and so on. Number of candidate paths can be changed flexibly.

4.1.2 Path Selector

The path selector selects optimal target path taking into account collision risk, navigation route, traffic rule, and riding comfort in the following equation.

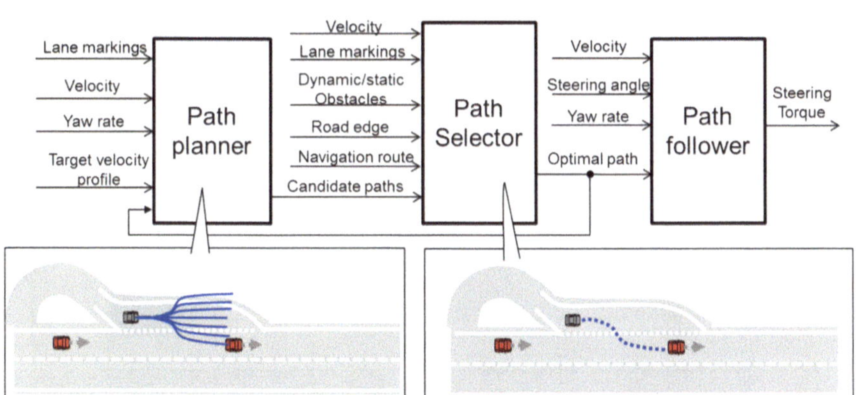

Fig. 14 Block diagram of path planning

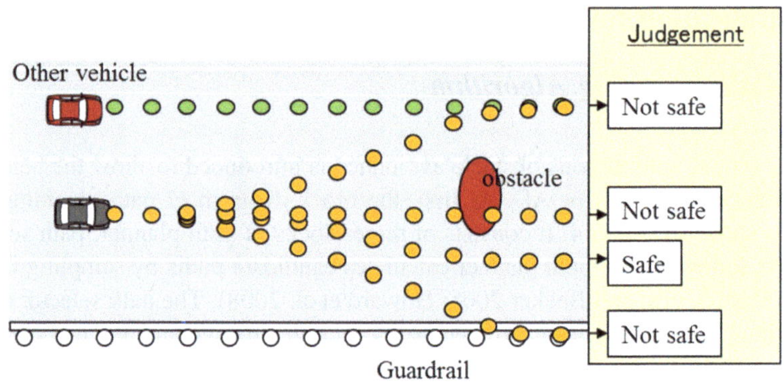

Fig. 15 Judgement for collision with obstacles

$$J = \int U_0 \mathrm{d}t + \int U_N \mathrm{d}t + \int U_{T1} \mathrm{d}t + \int U_{T2} \mathrm{d}t + \int U_R \mathrm{d}t \tag{1}$$

Here, U_0, U_N,U_{T1}, U_{T2}, U_R are cost function for collision risk with obstacle and road edge, cost function for following navigation route, cost function for prohibited traffic rule, cost function for traffic rule, and cost function for riding comfort, respectively. By comparing the candidate path with obstacles and road edge, the collision risk is calculated as shown in Fig. 15. When the path is judged as not safe, the penalty for collision risk is imposed.

Next, the priorities of the cost functions are explained. The collision risk is the most important, thus we put the highest priority on the collision risk. The prohibited traffic rule such as prohibited lane change is put the next high priority on. The following navigation route is put next high priority on. This means that if there is no collision and no prohibited traffic rule, the optimal path is selected based on the following navigation route. The next priority is traffic rule such as the vehicle normally does not drive on the lane marking and road shoulder. The lowest priority is riding comfort.

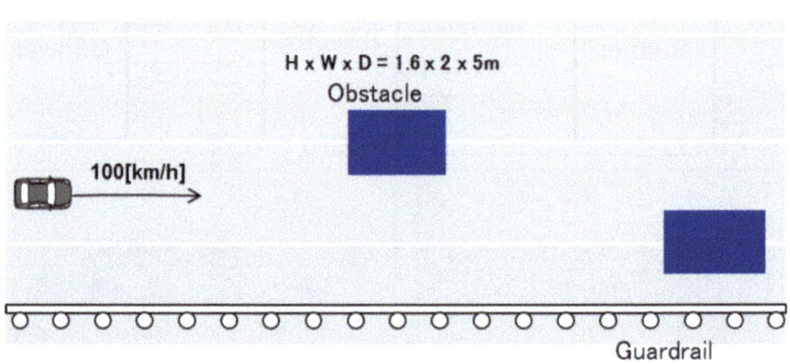

Fig. 16 Simulation condition for obstacle avoidance

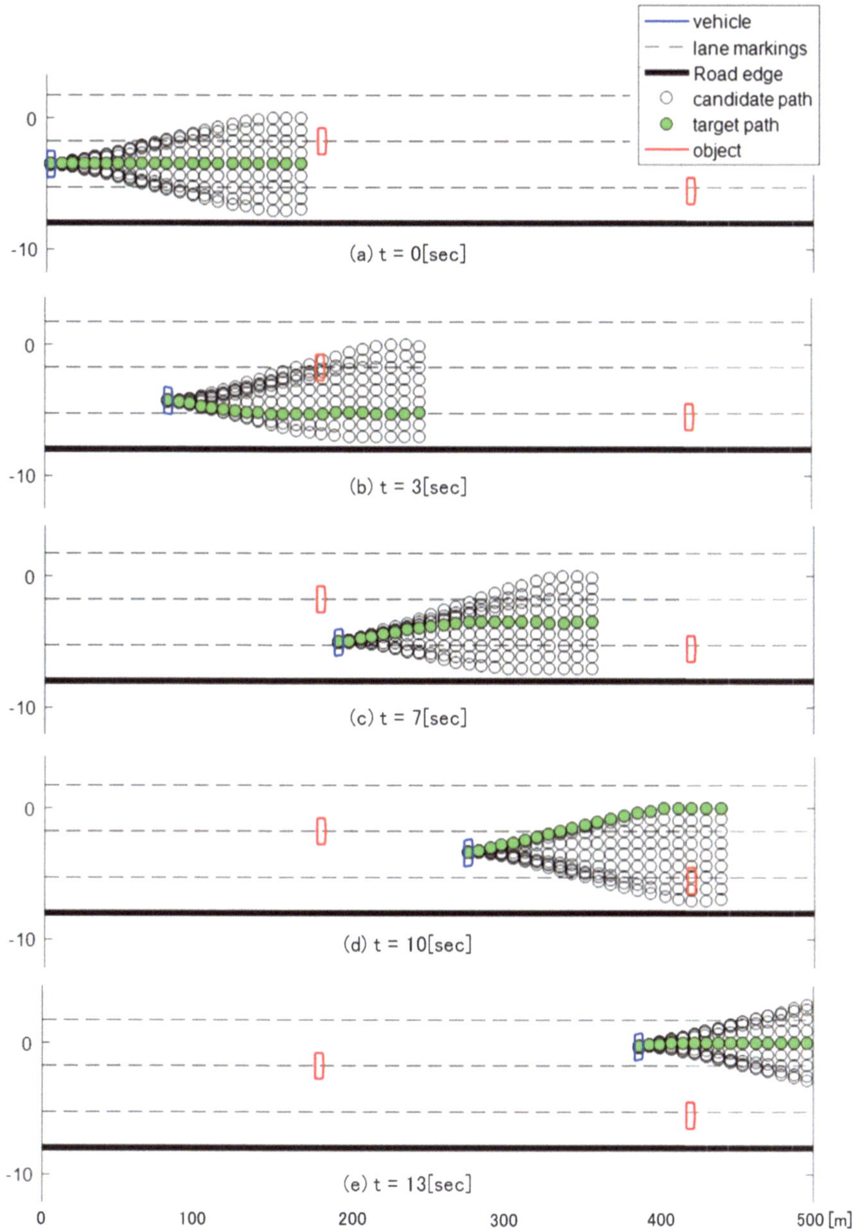

Fig. 17 Simulation result for obstacle avoidance: Candidate and optimal target path

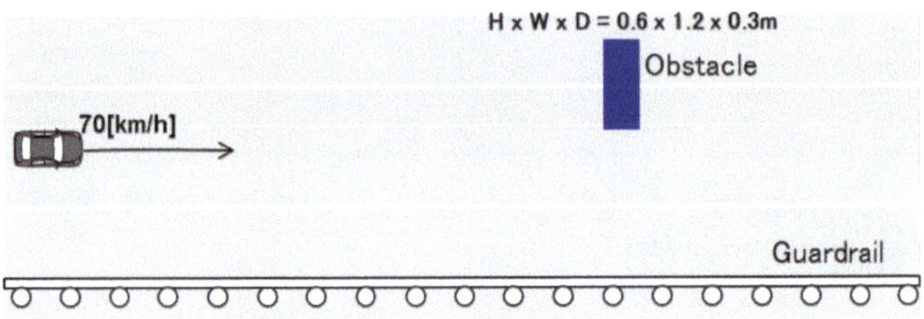

Fig. 18 Experimental condition for obstacle avoidance

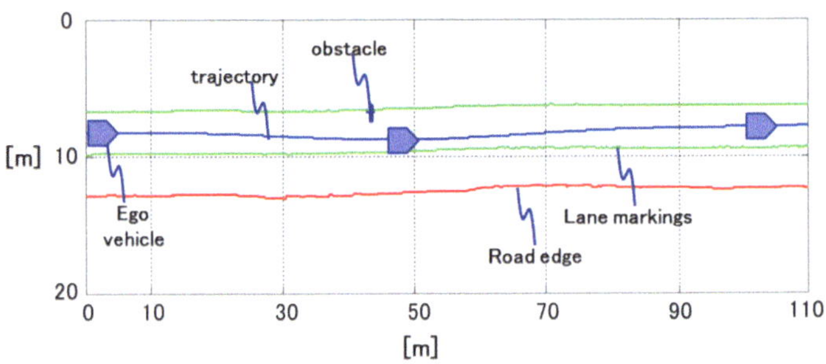

Fig. 19 Experiment result for obstacle avoidance: Trajectory of the vehicle and positions of related objects

4.2 Simulation Result

A simulation for the autonomous obstacle avoidance was conducted. The simulation condition is shown in Fig. 16.

A simulation result is shown in Fig. 17. In Fig. 17a, since there was no obstacle on the own lane up to end of the candidate paths, the algorithm selected the target path at the center of the own lane. In Fig. 17b and c, as the algorithm judged the road shoulder space as sufficient, the algorithm selected a maneuver to avoid obstacle through the road shoulder space. When the algorithm judged the road shoulder space as insufficient, the algorithm selected a lane change maneuver for obstacle avoidance as shown in Fig. 17d and e.

4.3 Experimental Result

An experiment for the autonomous obstacle avoidance was conducted. The experiment condition is shown in Fig. 18. In Fig. 19, a trajectory of the vehicle in the experiment is plotted based on GPS information. The position of road edge, lane marking, and obstacle are plotted based on the GPS information and the detection result by stereo camera. Figure 19 shows that the algorithm judged as the road shoulder space is sufficient and selected a maneuver to depart from the own lane for a moment to avoid obstacle.

5 Conclusion

As road edge information gives various benefits to ADAS/AD applications, we have developed the road edge detection algorithm with stereo camera in order to expand applicable scenarios. Two ADAS/AD applications using road edge information, validated by simulation and experiment, were introduced to show benefits of road edge information.

The integrated lateral assist system including LDP and RDP, which assumes lane support system in Euro NCAP, was proposed as an ADAS application. The coordination method of EPS and ESC for the integrated lateral assist system was proposed taking into account the characteristic of actuators and application. The experimental results showed that the proposed system realized LDP and RDP with stereo camera.

"Obstacle avoidance through road shoulder space" was introduced as an application for AD. The simulation and experimental result of autonomous obstacle avoidance showed that the system can judge road shoulder space as safe area with road edge information.

References

Dolgov D, Thrun S, Montemerlo M, Diebel J (2008) Practical search techniques in path planning for autonomous driving. In: Proceedings of the First International Symposium on Search Techniques in Artificial Intelligence and Robotics, June 2008

Howard MT, Green JC, Kelly A, Ferguson D (2008) State space sampling of feasible motions for high-performance mobile robot navigation in complex environments. J Field Robot

Kakegawa S, Sugawara T (2016) Road edge detection using dynamic programming with application to free space computation. In: 25th Aachen Colloquium Automobile and Engine Technology, p 1395–1418

Kuwata Y, Fiore G, Frazzoli E (2009) Real-time motion planning with applications to autonomous urban driving. IEEE Trans Control Syst Technol 17(5) (September)

Lattke B, Eckert A, et al (2015) Road departure protection-a means for increasing driving safety beyond road limits. In: 24th Enhanced Safety of Vehicles Conference 2015, vol 1, p 722–735

McNaughton M, Urmson C, Dolan J, Lee J (2011) Motion planning for autonomous driving with a conformal spatiotemporal lattice. In: IEEE International Conference on Robotics and Automation (ICRA)

Montemerlo M, Becker J et al Junior (2008) The stanford entry in the Urban challenge. J Field Robot

Nakada Y, Ota R, Kojima T, Takuya N (2011) Development of virtual lane guide. In: First International Symposium on Future Active Safety Technology toward Zero-Traffic-Accident

Williams A (2017) Euro NCAP 2018 5-Star Requirements: Asta Zero Testers'Day

Yamamura T (2008) Driver Assist System Considerations, First Human Factors Symposium: Naturalistic Driving Methods & Analyses
Yokoyama A, Saito S (2009) Development of the motorized direct yaw-moment control system: application to a lane-keep assist system(in Japanese). Int J Autom Eng Jpn

Robust and Numerically Efficient Estimation of Vehicle Mass and Road Grade

Paul Karoshi, Markus Ager, Martin Schabauer and Cornelia Lex

Abstract A recursive least squares (RLS) based observer for simultaneous estimation of vehicle mass and road grade, using longitudinal vehicle dynamics, is presented. In order to achieve robustness to unknown disturbances and varying parameters, depth is chosen in a sufficient way. This is done with a sensitivity analysis, identifying parameters with significant influence on the estimation result. The identification of vehicle parameters is presented in detail. The method is validated with an all-electric vehicle (AEV) using natural driving cycles. The results show little deviation between estimation and reference, as well as good convergence in urban areas, providing sufficient excitation. However, on highway roads, environmental influences like wind and slipstream of trucks, worsen the results, especially in combination with little excitation for the observer.

Keywords Mass estimation · Road grade estimation · Vehicle state estimation · Recursive least squares with forgetting

1 Introduction

The drivability of both passenger cars and commercial vehicles has significant dependence on the vehicle's mass and the road grade. Since the mass influences the vertical load on each wheel, longitudinal and lateral dynamics are influenced and

P. Karoshi (✉) · M. Ager · M. Schabauer · C. Lex
Institute of Automotive Engineering, Graz University of Technology, Inffeldgasse 11/2, 8010 Graz, Austria
e-mail: paul.karoshi@tugraz.at

M. Ager
e-mail: markus.ager@tugraz.at

M. Schabauer
e-mail: martin.schabauer@tugraz.at

C. Lex
e-mail: cornelia.lex@tugraz.at

C. Zachäus et al. (eds.), *Advanced Microsystems for Automotive Applications 2017*, Lecture Notes in Mobility, DOI 10.1007/978-3-319-66972-4_8

may cause safety-critical driving situations, such as braking or accelerating while cornering or on inhomogeneous road surface, (Heißing and Ersoy 2011; Kohlhuber and Lienkamp 2013). On the other hand, drivability may also include fuel consumption and emissions, (Heißing and Ersoy 2011). In this sense, especially those vehicles are affected that have a high variation of payload during operation. This is the case for commercial vehicles, but also for AEVs. Relative changes of physical parameters have big influence due to their lightweight design.

Automated driving functions that are already available in series production vehicles such as Adaptive Cruise Control (ACC) are designed in a way that they are robust to changes in vehicle mass due to payload and to road grade, since robust and reliable estimates are often not available. Some applications use rough estimates of the vehicle mass to adapt their intervention strategy, see, e.g., (Winner et al. 2009). With increasing level of automation, automated driving functions have additional requirements on detecting the vehicle's behavior, and thus increasing demands considering the accuracy of the road grade and the vehicle's mass. With SAE level of automation 3 and higher, responsibilities transfer step by step from the driver to the automated system (SAE International 2014). This also concerns legal considerations. It will no longer be the drivers' responsibility to, e.g., adapt speed and distance depending on the vehicle's expected behavior, but that of the system. To fulfill the requirements that come along with these responsibilities, robust and reliable estimates of both, vehicle mass and road grade are required.

Numerous studies on parameter estimation of land vehicles have been conducted in the past. An overview about applied methods is given in Rhode et al. (2015) and Kidambi et al. (2014).

Common methods for parallel mass and road grade estimation are Recursive Least Squares (RLS), Kalman Filters (KF), and Lyapunov theory based methods. These approaches are applied on models for longitudinal, lateral, and vertical vehicle dynamics and on their combination in multiple-model estimation. In Rozyn and Zhang (2010), the inertial parameters are identified by least squares analysis of the vehicle's sprung mass response, using a model for vertical vehicle dynamics. The approach is validated with an offline simulation. In Vahidi et al. (2005), an approach of RLS with multiple forgetting factors is proposed for a longitudinal vehicle model and is validated with experimental data of a truck, using additional sensor data. In McIntyre et al. (2009), a two-stage Ljapunov-based estimator for mass and road grade is proposed. In the first step, the constant vehicle mass and road grade are estimated. In the second step, the time-varying road grade is estimated with an observer. The approach is validated for a heavy duty vehicle. A short introduction to RLS Filtering and its derivation are given in Massachustetts Institute of Technology (2008).

In Mahyuddin (2014), an adaptive observer based approach with sliding mode term is presented. It uses a model for longitudinal vehicle dynamics and is validated online, on a small-scaled vehicle. In Kohlhuber and Lienkamp (2013), inertial parameters and various state variables are estimated with an unscented Kalman filter (UKF). This method is tested with driving maneuvers on an all-electric concept vehicle, using a single-track vehicle model. In Huh et al. (2007), a multiple-model approach is presented. RLS, KF, and dual-RLS are applied on a longitudinal, lateral, and vertical vehicle model. The algorithm is designed to estimate the vehicle mass in arbitrary driving situations. In Rhode et al. (2015), four approaches of Kalman Filters are compared with a longitudinal vehicle model. The method is validated online in a series production car. The multiple-model approach reaches best results in terms of robustness and wind-up stability. Detailed information on Kalman Filtering can be found in Grewal and Andrews (2001).

In this article, an estimation methodology is presented that simultaneously estimates the vehicle mass and the road grade using sensor data from the CAN-Bus of a series production vehicle. The goal is both to provide an easy implementation into a real-time system and to reach sufficient accuracy of the estimation results. Therefore, the algorithm is kept numerically simple compared to other approaches.

The measurement equipment, driving cycles, observer model and the observer using RLS with two different forgetting factors, are presented in Sect. 2. A parameter sensitivity analysis to investigate the accuracy of all needed vehicle parameters is conducted and the observer model is validated, both shown in Sect. 3. The methodology is validated using measurements from natural driving on urban roads, rural roads, and highways, see Sect. 4. Results of the presented methodology are discussed in Sect. 5, regarding expected accuracy in different scenarios using specific performance criteria.

2 Methodology

2.1 Test Vehicle and Test Tracks

In this study, the AEV Peugeot iOn is used as a test vehicle. It is equipped with an additional CAN-Bus logging device, a Global Navigation Satellite System (GNSS) sensor and an accelerometer. The primary goal of the method is to use only existing sensors of the car. So, all signals used in the proposed RLS algorithm are obtained from the CAN-Bus. Data from the GNSS and the accelerometer are used for validation of the longitudinal vehicle dynamics model and for calculating the reference road grade β_{ref}.

To test the method on real-world data, a natural driving cycle is chosen. It contains urban, highway, and rural roads. As a result, a wide range of driving situations is covered. While most of the urban part is on flat roads, the driving cycle also includes elevation changes on highway and rural roads of up to $\beta = \pm 0.1$ rad. Before each run, the car is weighted on scales to consider the different number of passengers. Table 1 gives an overview on the road segments of the test track.

Table 1 Distribution of road segments of the test track

Road	Distance (km)	Percentage (%)
Urban	9.6	30.00
Highway	14.0	43.75
Rural	8.4	26.25
Total	32.0	100.00

2.2 System Model

Within this approach, only longitudinal vehicle dynamics is considered. Figure 1 shows the forces acting on the vehicle in longitudinal direction.

The equation of longitudinal motion reads:

$$m \cdot \dot{v} = F_{\text{long}} - F_{\text{aero}} - F_{\text{climb}} - F_{\text{roll}} - F_{\text{brake}}, \tag{1}$$

with

$$F_{long} = \frac{T \cdot i \cdot \eta}{r_e} - \frac{J \cdot \dot{v}}{r_e^2}, \tag{2}$$

$$F_{aero} = \frac{1}{2} \cdot \rho \cdot c_W A \cdot v^2, \tag{3}$$

$$F_{\text{climb}} = m \cdot g \cdot \sin \beta, \tag{4}$$

$$F_{\text{roll}} = m \cdot g \cdot c_r \cdot \cos \beta, \tag{5}$$

$$F_{brake} = c_b \cdot p_b. \tag{6}$$

The vehicle's longitudinal velocity v is calculated from the rotational speed of left and right front wheels: $v = (\omega_l + \omega_r)/2$. This is a feasible approach, because the vehicle is rear-wheel-driven and the mass estimation is paused while braking. Thus, the error due to possible tire slip is small. The vehicle's acceleration $\dot{v}$ is calculated by differentiation of v. In Eqs. (1)–(6), T denotes the electric motor driving torque, i the gearbox ratio, η the gearbox efficiency, r_e the effective tire radius, J the drivetrain's moment of inertia, ρ the air density, $c_W A$ the combination of air drag coefficient and the cross-sectional area of the vehicle, m the vehicle mass, g the gravitational constant, β the road grade, and c_r names the rolling resistance.

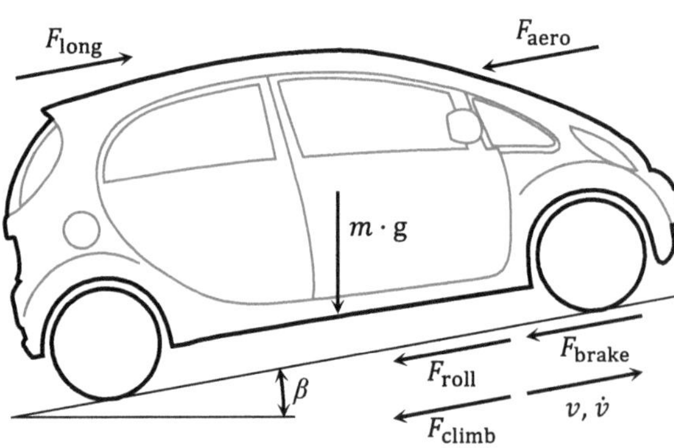

Fig. 1 Longitudinal forces acting on the vehicle

Within this simple approach, longitudinal slip between tire and road is neglected. Therefore, no explicit tire model is implemented. Furthermore, the effective tire radius r_e is assumed to be constant. For small β, $\sin\beta \sim \beta$ and $\cos\beta \sim 1$ are assumed. The brake force F_{brake} depends on the brake pressure p_{b}, geometric properties of the brake and on the friction coefficient between brake disk and brake pad. Within this approach, these influences are summarized in one parameter c_{b}.

Equation (1) can be transformed to the linear equation of the system model:

$$y = \boldsymbol{\Phi}^{\mathrm{T}} \cdot \boldsymbol{\theta}, \tag{7}$$

where y denotes the vehicle's acceleration, $\boldsymbol{\Phi}$ is the system vector and $\boldsymbol{\theta}$ indicates the vector of estimated parameters:

$$y = \dot{v}, \tag{8}$$

$$\boldsymbol{\Phi}^{\mathrm{T}} = \left[F_{\mathrm{long}} - F_{\mathrm{aero}} - F_{\mathrm{brake}}, -g\right], \tag{9}$$

$$\boldsymbol{\theta} = \left[\frac{1}{m}, \beta + c_{\mathrm{r}}\right]. \tag{10}$$

On the vehicle's CAN-Bus, signals for ω_{l}, ω_{r}, T, and p_{b} are available. The signals are filtered with a low-pass filter, in order to compute a smooth signal for $\dot{v}$.

The identification of the constant vehicle parameters c_{b}, c_{r}, $c_{\mathrm{W}}A$ and r_{e} is discussed in Sect. 3.2.

2.3 Recursive Least Squares (RLS) Algorithm

The RLS algorithm is a well-known approach for estimation of constant parameters. For the estimation of time-varying parameters, a forgetting factor can be introduced, see Massachustetts Institute of Technology (2008), Vahidi et al. (2005). RLS with exponential forgetting minimizes the total squared error ε in the sense of least squares: $\varepsilon = \sum_{i=1}^{k} \lambda^{k-i}\left(y_i - \boldsymbol{\Phi}_i^{\mathrm{T}} \cdot \hat{\boldsymbol{\theta}}_k\right)^2$, where λ denotes a scalar forgetting factor, with $0 < \lambda \leq 1$. Old data is continuously less weighted, than new data. Thus, time-varying parameters can be tracked. If parameters change at different rates, (Vahidi et al. 2005) propose an approach with multiple forgetting factors. Then λ becomes $\lambda = diag[\lambda_1, \lambda_2]$, with λ_1 and λ_2 as the forgetting factors for the estimated parameters $\hat{\boldsymbol{\theta}}_1$ and $\hat{\boldsymbol{\theta}}_2$. Since the mass is constant, its forgetting factor is chosen to be close to unity. For time-varying parameters values between $0.95 < \lambda \leq 0.995$ are found in literature (Massachustetts Institute of Technology 2008). The chosen values for the forgetting factors are $\lambda = [10.98]$.

For online applications, the recursive scheme of an RLS with multiple forgetting factors reads (Massachustetts Institute of Technology 2008; Vahidi et al. 2005):

(1.) For each timestep k, the filter output is given by:

$$y_k = \boldsymbol{\Phi}_k^{\mathrm{T}} \cdot \hat{\theta}_{k-1}. \tag{11}$$

(2.) The estimation error is:

$$e_k = \dot{v}_k - y_k. \tag{12}$$

(3.) Then the Kalman-Gain is:

$$\boldsymbol{l}_k = \frac{\boldsymbol{P}_{k-1}\boldsymbol{\Phi}_k}{1 + \boldsymbol{\Phi}_k^T \boldsymbol{P}_{k-1}\boldsymbol{\Phi}_k} \tag{13}$$

(4.) The updated covariance matrix $\boldsymbol{P}_k$ reads:

$$\boldsymbol{P}_k = \lambda^{-1}\left[\boldsymbol{P}_{k-1} - \boldsymbol{l}_k \boldsymbol{\Phi}_k^T \boldsymbol{P}_{k-1}\right]\lambda^{-1}, \tag{14}$$

(5.) Finally, the parameters are updated:

$$\hat{\theta}_k = \hat{\theta}_{k-1} + \boldsymbol{l}_k e_k \tag{15}$$

Forgetting is a widely used approach in online parameter estimation. However, difficulties can occur during periods with low excitation. Then, old information is “forgotten”, while the new information has a low signal-to-noise ratio. This leads to a “wind-up” of the covariance matrix $\boldsymbol{P}_k$ and instable estimation results (Fortescue et al. 1981). Therefor two solutions are applied:

(1) The mass estimation is paused, if either the acceleration is small, $|\dot{v}| < 0.6\,\frac{\mathrm{m}}{\mathrm{s}^2}$, or the mechanical friction brake is applied, $p_{\mathrm{b}} \neq 0$.
(2) The estimator output for the vehicle mass is limited with minimum and maximum values, $\frac{1}{m_{\max}} < \frac{1}{m} < \frac{1}{m_{\min}}$.

Filters are often initialized with the expectations of the estimated parameters, (Bauer 2007). In vehicle dynamics, controllers rely on the estimated vehicle mass, such as the calculation of electric driving range in AEVs and hybrid electric vehicles (HEVs) or load-dependent brake pressure and gearshift strategies in commercial vehicles. To ensure system availability and to avoid dangerous driving states, the estimation is initialized with the maximum vehicle mass and a road grade of zero:

$$\hat{\boldsymbol{\theta}}_0 = \left[\frac{1}{m_{\max}}, c_{\mathrm{r}}\right]^{\mathrm{T}}. \tag{16}$$

To ensure that $\boldsymbol{P}_k$ does not become singular for small k, it is common to define $\boldsymbol{P}_0 = \delta\mathbf{I}$, with $\delta > 100\sigma_{\Phi}^2$, (Massachustetts Institute of Technology 2008), where σ_{Φ}^2 is the variance of the input. In this approach, $\boldsymbol{P}_k$ is initialized with

$$\boldsymbol{P}_0 = \mathrm{diag}[0.1, 0.5]. \tag{17}$$

Properties of stability and convergence of RLS with forgetting are discussed in Vahidi et al. (2005). Only little information about stability, convergence, and tracking capabilities in time-varying systems can be found, (Vahidi et al. 2005). A proof of convergence for constant parameters is found in Parkum et al. (1992).

The observer is modeled in MATLAB/Simulink®. This brings the advantage that the model can easily be compiled in order to run on real-time hardware. However, this study will deal with real-time tests on a later occasion.

3 Sensitivity Analysis and Parameter Estimation

3.1 Sensitivity Analysis

A sensitivity analysis is carried out, in order to identify parameters with significant influence on the estimation results. The analysis is performed with the built-in sensitivity analysis tool in MATLAB/Simulink®. Five parameters are chosen: Rolling resistance c_{r}, gear ratio i, air density ρ, tire radius r_{e}, and the air drag coefficient $c_{\mathrm{W}}A$. In the first step, the parameters indicated by x are varied by $\pm$ 10% of their initial value and the cost functional y, here the estimated mass m is evaluated. The partial correlation coefficients R_{ij}, describing the correlation between a parameter x and the cost functional y, are:

$$R_{i,j} = \frac{C_{i,j}}{\sqrt{C_{i,i}C_{j,j}}}, \tag{18}$$

with $C_{ij} = E\left[(x - \mu_x)(y - \mu_y)\right]$, $\mu_x = E[x]$ and $\mu_y = E[y]$.

Figure 2 depicts the linear partial correlation R. The results indicate the parameter's influence on the estimated mass. The most sensitive parameter for mass estimation is the effective tire radius r_{e}. Other important parameters are the moment of inertia J and the rolling resistance c_{r}. The unknown J corresponds with r_{e}, Eq. (2). Therefore, these parameters are correlated. Careful identification of these parameters will result in improved quality of the estimation results. The air density ρ and the air drag $c_{\mathrm{W}}A$ show less correlation with the estimated mass. However,

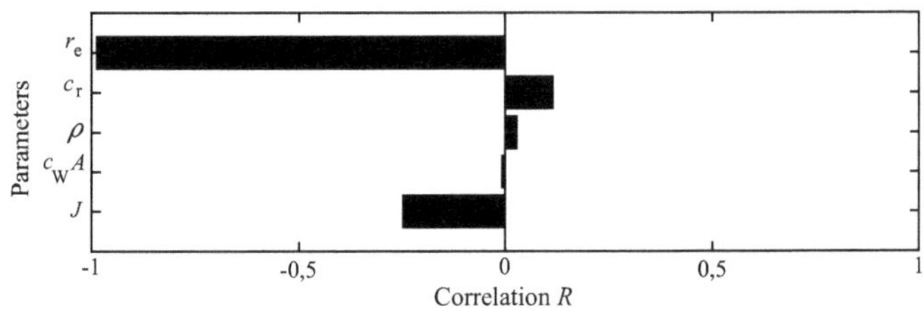

Fig. 2 Linear correlation R of parameters and estimator output

$c_W A$ has significant influence on vehicle dynamics at higher velocities, Eq. (3). So a good estimate of this parameter is desirable.

3.2 *Identification of Parameters and Validation of the Vehicle Model*

For the following parameters, no information was available. They were either measured or identified offline from measurement data.

As the sensitivity analysis shows, the effective tire radius r_e significantly influences the mass estimation, since it is the only and important parameter regarding the description of longitudinal tire force transmission within this approach. Therefore, it has to be determined sufficiently accurate. The effective rolling radius of a tire depends especially on the tire load, the inflation pressure, the vertical tire stiffness, respectively, and the tire speed, e.g., (Hirschberg and Waser 2012; Rill 2011). In a simple approach, it can be determined by measuring the total traveled distance (effective rolling circumference) and the total angle of rotation of a tire under free rolling conditions at a very low velocity. By such measurements at different tire loads, an averaged effective radius of $r_e = 0.277\text{m}$ for the tires of the test vehicle was obtained. This value also corresponds well to measurements of these tires which were carried out on a flat track tire test rig at a constant speed of 16.7m/s.

The rolling resistance c_r and the air drag coefficient $c_W A$ are identified offline with measurement data from steady-state driving maneuvers on flat road, with three different payloads. Figure 3a depicts the velocity and vehicle mass of the selected data points. At constant vehicle velocity on flat road, Eqs. (1)–(6) simplify to:

$$0 = \frac{T \cdot i \cdot \eta}{r_e} - \frac{1}{2} \cdot \rho \cdot c_W A \cdot v^2 - m \cdot g \cdot c_r. \tag{19}$$

In this equation, the moment of inertia J disappears, and a linear system $y = \boldsymbol{\Phi}^T \cdot \boldsymbol{\theta}$ can be defined as:

$$y = \frac{T \cdot i \cdot \eta}{r_e}, \tag{20}$$

$$\Phi^T = \left[\frac{1}{2} \cdot \rho \cdot v^2, m \cdot g\right], \tag{21}$$

$$\theta = [c_W A, c_r]. \tag{22}$$

Here, y, T, v and m are vectors of size nx 1, with n representing the number of data points. This system is over determined. It is solved in the sense of minimal least squares, with the pseudoinverse of Φ being

$$\hat{\theta} = \left(\Phi^T \Phi\right)^{-1} \Phi^T y. \tag{23}$$

Once the values of c_r and $c_W A$ are identified, an analogue approach is applied for the identification of the moment of inertia, J. Again, data are taken from test drives on a flat road. It is intuitive to select maneuvers where J has significant influence on the vehicle dynamics. Considering Eq. (2), this is the case at high accelerations. A roll out test has the advantage that the tire slip is small, the driving torque T is zero and no brakes are applied. However, the deceleration $\dot{v}$ is small and no reliable results were achieved with this maneuver. Therefore, events of acceleration were selected for the identification of J. Figure 3b depicts the velocity and vehicle mass of the selected data points.

The linear system $y = \Phi \cdot \theta$ can now be written as:

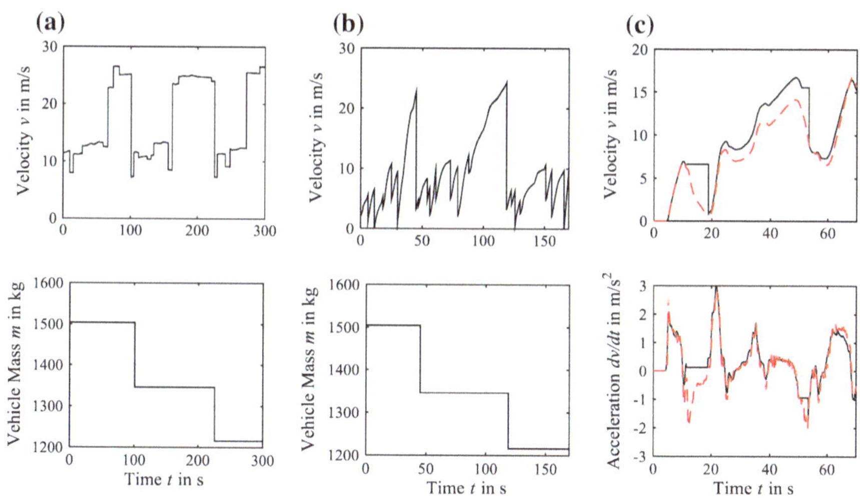

Fig. 3 Data of v and m for identification of $c_W A$ and c_r (**a**) and J (**b**), validation results (**c**). Reference in *dashed*- and estimated data in *solid lines*

$$\boldsymbol{y} = m \cdot \dot{\boldsymbol{v}} - \frac{\boldsymbol{T} \cdot i \cdot \eta}{r_{\mathrm{e}}} + \frac{1}{2} \cdot \rho \cdot c_{\mathrm{W}} A \cdot \boldsymbol{v}^2 + \boldsymbol{m} \cdot g \cdot c_{\mathrm{r}}, \tag{24}$$

$$\boldsymbol{\Phi} = -\frac{\dot{\boldsymbol{v}}}{r_{\mathrm{e}}^2}, \tag{25}$$

$$\theta = J. \tag{26}$$

Again, $\boldsymbol{y}$, $\boldsymbol{T}$, $\boldsymbol{v}$ and $\boldsymbol{m}$ are vectors of size nx 1, with n representing the number of data points. This over determined system is solved in the sense of minimal least squares. The resulting identified values are:

$$c_{\mathrm{r}} = 0.0099, c_{\mathrm{W}}A = 0.7479\mathrm{N}^2\mathrm{s}^4, J = 6.65\mathrm{kg\ m}^2$$

These values are in good coincidence with values known from literature (Rill 2011). The mathematical model from Eq. (1) and the identified parameters are validated in MATLAB/Simulink®, with measurement data. Inputs to the model are the measured driving torque T and the brake pressure p_{b}. The road grade β_{ref} is calculated from measured GNSS coordinates and map data. Thus, the simulation results may diverge at some points, where the map data does not accurately represent the real road altitude. The simulation is paused whenever the mechanical friction brake is applied. At the end of each brake application, the velocity is reset to the measured value. Figure 3c shows the simulation results. The simulated acceleration shows good coincidence with the measured signal. One can identify the periods of brake application. The simulated velocity shows good coincidence with the measured values too. However, at some points, the integration of errors in the simulated acceleration leads to an offset of the velocity signal. Figure 4 exemplarily depicts the error e of the simulated acceleration against the driving torque T, velocity v, and acceleration $\dot{v}$.

Apparently, there are no clear correlations between these quantities. Therefore, the remaining error is caused by inaccuracies in the calculated road grade, wind, or vehicle dynamics not covered by the model.

4 Results

For evaluation of the observer's results, three performance criteria are defined:

- t_{c}: Time to convergence until the estimated mass has converged to 5% of its final value.
- e_{m}: Root mean squared error of estimated mass m.
- e_{β}: Root mean squared error of estimated road grade β.

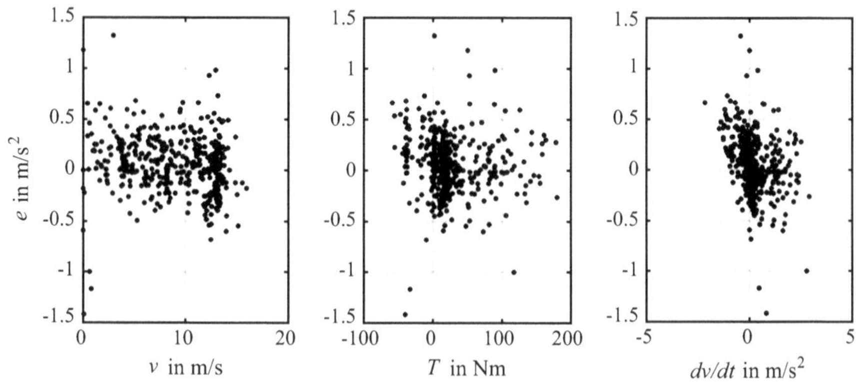

Fig. 4 Comparison of estimation error to v, T and $\dot{v}$

4.1 Validation with a Numerical Model

As a test for stability and convergence, the observer's performance is validated with a numerical model, representing Eqs. (1)–(6). In this model, the driving torque T is a sine function and the road grade is zero. Figure 5 depicts the validation results. It can be seen that the estimator runs into its limits at the estimation start. The RMS-error of m is $e_{\mathrm{m}} = 2.8\%$, the RMS-error of β is $e_{\beta} = 0.0034$rad, and the time to convergence is 27.8s.

4.2 Results in Real-World Driving Conditions

On the test track, five driving tests were performed with varying vehicle mass m from 1216kg with one passenger to 1504kg with four passengers.

Offline simulations of the estimator are started at the beginning of urban, highway, and rural parts, respectively. Therefore, results of 15 estimations are available.

Figure 6 exemplarily shows the observer's performance. The measured vehicle mass is $m = 1346$kg, $t_{\mathrm{c}} = 89.4$s, $e_{\mathrm{m}} = 4.8\%$ and $e_{\beta} = 0.025$ rad. The estimated road grade β matches well with its reference β_{ref}. However, β_{ref} is calculated from map data and may contain outliers. Therefore, the results of e_{β} may not be accurate, but deliver a reasonable reference. Finally, the altitude trajectory, calculated from β, is depicted. Although there is a drift, characteristics of the reference are met well.

Table 2 lists the average estimation results, categorized by road type.

It is apparent that the best results are achieved on urban roads, followed by rural and highway roads. This is mainly caused by dynamic driving in the urban area, with high excitation by acceleration events. On rural roads, there are more parts with constant driving, therefore less excitation. On highway roads not so good

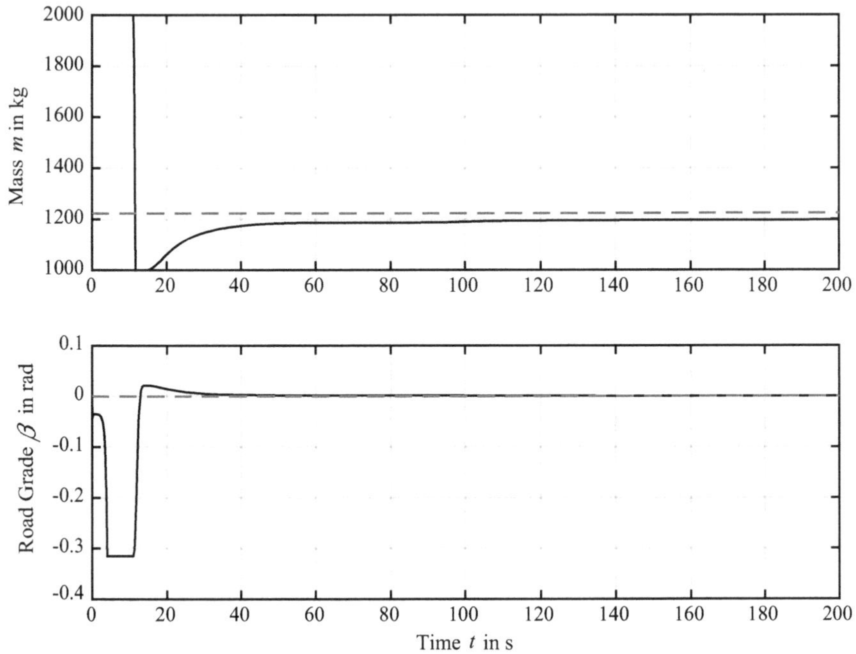

Fig. 5 Validation results with a numerical model. Reference in *dashed*- and estimated data in *solid lines*

results are achieved, usually due to the little excitation of just one acceleration event on the on-ramp. Additionally, on a highway significant disturbances, such as slipstream of trucks, and wind can occur, which are not covered by the observer model.

5 Summary

Mass and road grade have significant influence on the drivability of vehicles, e.g., commercial vehicles during gear shift, as well as on fuel consumption and emissions of HEVs and vehicle dynamics in general. In this article, an RLS-based observer is presented that simultaneously estimates these parameters. The observer considers the longitudinal vehicle dynamics and only uses data usually available on the on-board CAN-Bus. In order to account for the parameter's different rates of change, two forgetting factors are applied to the RLS. An easy implementation into a real-time system is aimed for, thus the algorithm is kept numerically simple. In a parameter sensitivity analysis, the accuracy of various needed vehicle parameters is investigated. Issues like fast convergence, stability, and persistent excitation are discussed.

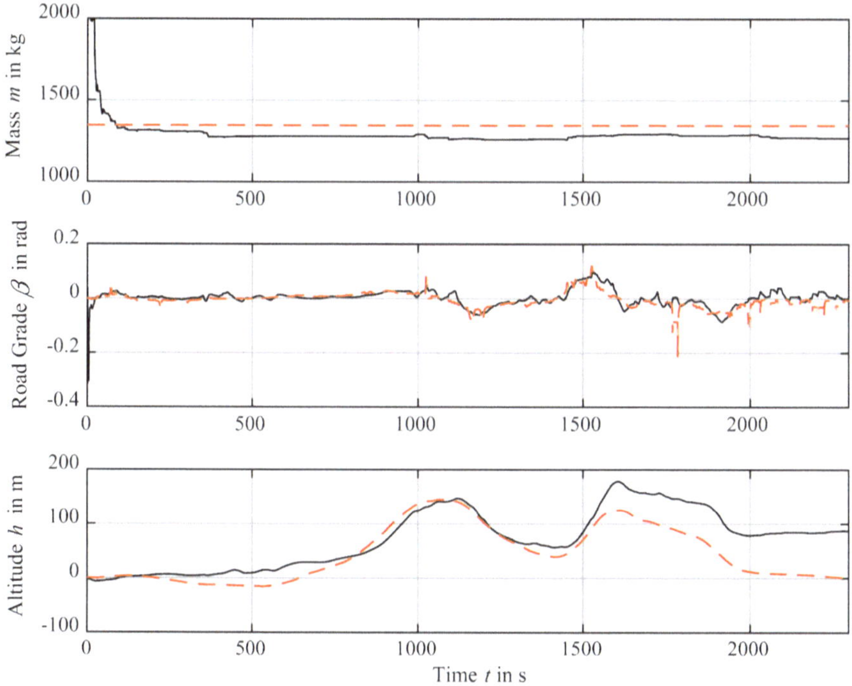

Fig. 6 Observer performance with natural driving data. Reference in *dashed*- and estimated data in *solid lines*

Table 2 Estimation results, categorized by road type

Criteria	Urban	Highway	Rural
t_c	101.7 s	367 s	824 s
e_m	4.6%	14.3%	9.6%
e_β	0.039 rad	0.11 rad	0.08 rad

The results are validated with test drives, carried out on urban, rural, and highway roads to demonstrate the robustness. Estimation results from different driving states and environmental conditions are presented. The results show little errors in estimation and good convergence in urban areas, where sufficient excitation is provided. These results match with results from approaches found in literature. However, on highway roads, environmental influences worsen the results, like wind and slipstream of trucks in combination with little excitation of the observer.

References

Bauer R (2007) Zustandschätzung und Filterung. Institut für Regelungs- und Automatisierungstechnik, Graz Technical University, Textbook

Fortescue TR, Kershenbaum LS, Ydstie BE (1981) Implementation of self-tuning regulators with variable forgetting factors. Automatica 17:831–835

Grewal MS, Andrews AP (2001) Kalman filtering: theory and practice using MATLAB. Wiley, New York

Heißing B, Ersoy M (2011) Chassis handbook-fundamentals, driving dynamics, components, mechatronics, perspectives. Springer, Wiesbaden, Germany

Hirschberg W, Waser HM (2012) Kraftfahrzeugtechnik. Institute of Automotive Engineering, Graz Technical University, Textbook

Huh K, Lim S, Jung J, Hong D, Han S, Han K (2007) Vehicle mass estimator for adaptive roll stability control. SAE World Congress

Kidambi N, Harne RL, Fujii Y, Pietron GM, Wang KW (2014) Methods in vehicle mass and road grade estimation. SAE Int Passeng Cars Mech Syst 7

Kohlhuber F, Lienkamp M (2013) Online estimation of physical vehicle parameters with ESC sensors for adaptive vehicle dynamics controllers 13. Internationales Stuttgarter Symposium Automobil-und Motorentechnik, 157–175

Mahyuddin MN (2014) Adaptive observer-based parameter estimation with application to road gradient and vehicle mass estimation. IEEE Trans Ind Electron

Massachustetts Institute of Technology (2008) Signal processing: continuous and discrete, Introduction to Recursive-Least-Square (RLS) adaptive filters. MIT OpenCourseWare. http://mit.edu

McIntyre ML, Ghotikar T, Vahidi A, Song X, Dawson DM (2009) A two-stage lyapunov-based estimator for estimation of vehicle mass and road grade. IEEE Trans Veh Technol 58:3177–3185

Parkum JE, Poulsen NK, Holst J (1992) Recursive forgetting algorithms. Int J Control 109–128

Rhode S, Hong S, Hedrick J, Gauterin F (2015) Vehicle tractive force prediction with robust and windup-stable Kalman filters. Control Eng Pract 46:37–50

Rill G (2011) Road vehicle dynamics-fundamentals and modelling, 1st edn. CRC Press-Taylor and Francis Group, p 61

Rozyn M, Zhang N (2010) A method for estimation of vehicle inertial parameters. Veh Syst Dyn 48

SAE International (2014) SAE J3016: Taxonomy and definitions for terms related to on-road motor vehicle automated driving systems. Standard

Vahidi A, Stefanopoulou A, Peng H (2005) Recursive least squares with forgetting for online estimation of vehicle mass and road grade: theory and experiments. Int J Veh Mech Mobil 45:31–55

Winner H, Hakuli S, Wolf G (2009) Handbuch Fahrerassistenzsysteme Grundlagen, Komponenten und Systeme für aktive Sicherheit und Komfort, 1st edn. Vieweg + Teubner, Wiesbaden

Fast and Accurate Vanishing Point Estimation on Structured Roads

Thomas Werner and Stefan Eickeler

Abstract We propose a method for estimating the vanishing point of structured roads directly in the image plane using the parallel nature of road markings as well as intelligent preprocessing and data reduction steps. The resulting vanishing point enables estimating the image to world projection, which then is used to perform subsequent tasks such as object detection. The major advantages of the proposed method are modest computational requirements as well as independence of the used camera model and without a calibration phase.

Keywords Fast vanishing point estimation · Real time · Camera independent · Automotive · Inverse perspective mapping · Bird's eye view · Structured roads · Road marking · Multiresolution box filter

1 Introduction

An important aspect of many automotive-related computer vision problems is road modeling, which is the basis for many other tasks regarding driver-assistance systems like lane or vehicle detection. For this, in-car camera systems most often work with a single lens camera as the input source. Based on a single image, the bird's eye view, a projection between image and road plane, is computed (Bertozz et al. 1998). Using this projection, every image pixel is mapped to a real-world location on the road, which makes it possible to measure distances as well as angles between remapped objects. These measurements are the basis for a range of detection algorithms, e.g., lane marking detection (Borkar et al. 2009; Deng 2013) or parking assistants (Lin and Wang 2010).

T. Werner (✉) · S. Eickeler
Fraunhofer Institute for Intelligent Analysis and Information Systems, 53757 Schloss Birlinghoven, Sankt Augustin, Germany
e-mail: thomas.werner@iais.fraunhofer.de

S. Eickeler
e-mail: stefan.eickeler@iais.fraunhofer.de

C. Zachäus et al. (eds.), *Advanced Microsystems for Automotive Applications 2017*, Lecture Notes in Mobility, DOI 10.1007/978-3-319-66972-4_9

General drawbacks of these methods are the reliance on knowledge of the camera model as well as an initial calibration phase. Additionally, one-time calibration does not allow any adaptation to certain road characteristics like slopes. Furthermore, the complicated computations involved make this approach hardly usable for real-time processing, because of the significant resources of the limited hardware needed.

An alternative way of approximating the inverse perspective mapping (IPM) between image and road plane is to use the vanishing point and a pinhole camera model. This way, it is possible to estimate the IPM in a very fast way without any calibration. Furthermore, if the vanishing point is iteratively estimated, it is possible to adapt to changes of the ground, e.g., slope changes.

In this paper, we present a method for continuously estimating the vanishing point within a video of structured roads with very high speed based on simple assumptions about the surrounding and the used camera model. Exploiting appearance priors of road markings, e.g., symmetry and contrast, makes the proposed algorithms core already computationally cheap, the efficient preprocessing and various data reduction steps give the biggest advantage over other real-time vanishing point estimation algorithms.

2 Vanishing Point

A vanishing point is a point in an image where the projection of parallel lines (in world space) intersects each other due to the perspective of the camera. Although an image can contain up to three such points, one for each real-world dimension, in the context of in-car camera systems the most interesting setting is the one-point perspective. This perspective characteristically consists of lines perpendicular to the image plane with all pairs of parallel lines converging to a single vanishing point on the horizon line. Figure 1 shows an example of a one-point perspective.

3 System Overview

This section describes the process from the input frame to the final estimated vanishing point. Figure 2 shows an overview of the single stages.

3.1 Double-Edge Detection

The first step of the proposed algorithm is lane marking detection. The basic assumption is that lane markings are bright objects on a darker ground. From this, two constraints can be derived. First, we use a simple brightness threshold to concentrate on the portions of the image that could be markings; second, we assume

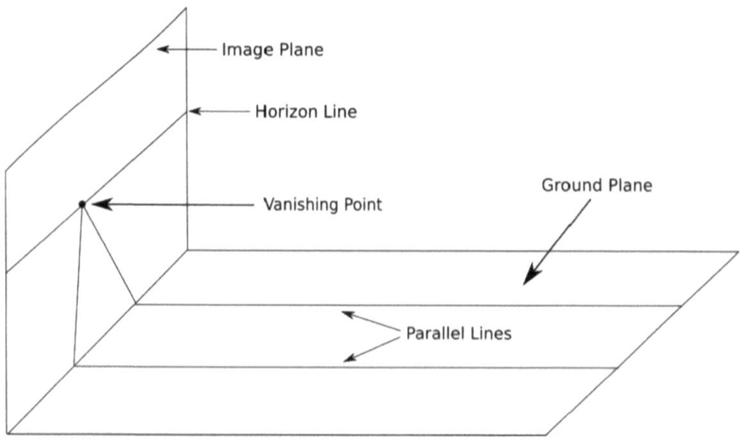

Fig. 1 Visualization of the one-point perspective and the corresponding vanishing point on the *horizon line*

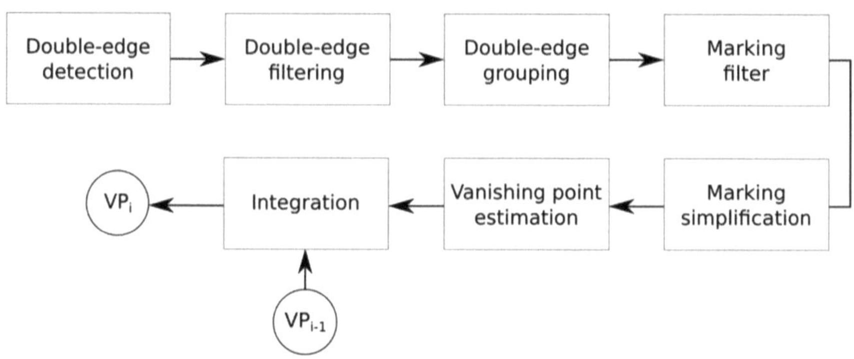

Fig. 2 Overview of the proposed method

that a marking border consists of double-edge pairs. A double-edge pair is a pair of two edges on one image line, where one edge transitions from dark to light (min-edge) and the other from light to dark (max-edge). In addition to that, it can be safely assumed that the min- and max-pixel of a pair cannot be arbitrarily far apart or close together. We call the pixels between a double-edge pair as a double-edge segment.

Given an image of a road scene, our approach processes it in a highly efficient two-pass line-by-line manner. To be more precise, the algorithm keeps a buffer of three image lines, which contain the current line as well as the ones visited previously and visited next. Based on this buffer, our approach is able to perform Gaussian smoothing as well as to construct an integral image line of the current line in a single pass. The second pass over the line is used to detect all min- and max-edges along it by applying a one-dimensional multiresolution box filter that was designed to detect edges of various sizes and strength. Figure 3 shows the used filter.

During the second pass, the algorithm constantly keeps track of edge pixels visited before and matches pairs of consecutive min- and max-edges. In the next step, the detected double-edges are filtered to reduce the input for all subsequent steps.

3.2 Double-Edge Filtering

After a double-edge pair is detected, the aforementioned constraints about intensity and distance can be applied to either accept or reject the detection. First, pixels between both edges must have a minimum intensity value to be accepted as a marking. However, during our experiments, we encountered different situations where even brighter visible object were falsely accepted. Most often this was due to glaring lights either directly shining into the camera, e.g., breaking lights or front lights of approaching traffic at night, or getting reflected on wet surfaces. Thus, we decided to set the minimal intensity threshold to a value of 50 to also account for yellow as well as dirty marking, and the maximal threshold to be 250, to exclude very bright spots and glares.

Second, the distance between both edges cannot be too large or too small. Implied by the perspective of the scene, it is clear that those distance thresholds vary depending on the vertical distance to the vanishing point. Since marking far from the camera appear to be smaller, the selected threshold should equally be dependent on the vertical position. To keep the algorithm as efficient as possible, we perform a linear interpolation between two sets of width thresholds, one for the bottommost line and one for the line where the vanishing point was detected in the

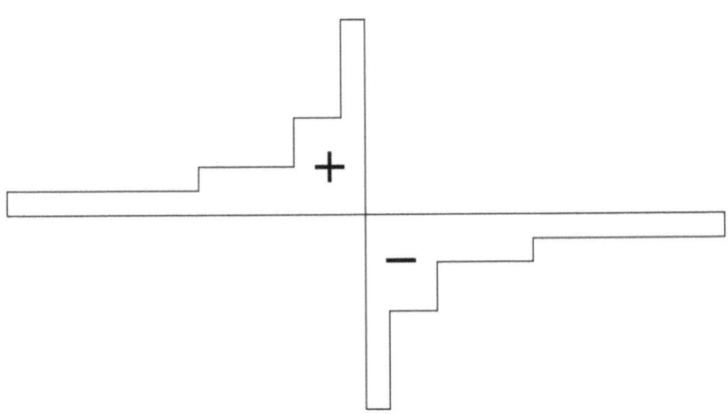

Fig. 3 Multiresolution box filter used in the method. It can be efficiently computed given an integral image. The advantage is that edges of varying sizes and strength can be detected

previous frame. Each set consists of a minimum and maximum allowed width value. For the bottommost line, those thresholds are set to be 10 and 100 pixels and at the horizon line, those are both 0. We have found that this gives reasonable results for an image with a width of 1280 pixels.

Figure 4 shows the results after the first two steps on a test image. It can be seen that the two primary lane markings are detected reliably, whereas no parts of the asphalt are considered as markings. However, since we only use soft constraints, we also get a number of false positive detections, mainly produced by high-frequency regions, like the leaves on the side, or structures similar to road markings, like the reflector post.

3.3 *Double-Edge Grouping to Lane Markings*

This step is concerned with the grouping of double-edges into coherent groups that together form a part of a lane marking.

To achieve this, all surviving double-edges are processed a second time from bottom to top and linked with each other based on yet another set of constraints. A double-edge is expected to appear in the next line near the position of the current one, since lane marking are closed regions. Additionally for the same reason, the width of both double-edge segments should be almost the same.

Both constraints are formulated in a straightforward way: The vertical distance constraint can be checked by looking at the distance between the two lines the double-edges appear in. We introduced a threshold parameter, by default set to 10, to filter out segments accordingly. The horizontal distance and the size constraint can be checked by a single measure: the Jaccard coefficient, also known as Intersection over Union or IoU. This is the coefficient of the intersecting area over the union of both areas. The range of the IoU is [0, 1], where 0 is assigned to objects that do not intersect at all, whereas 1 indicates identical objects. Since we are interested in the horizontal offset and the width difference and expect double-edges

Fig. 4 Results of the double-edge detection phase. It can be seen that the relevant lane markings are detected robustly. However, we have some false detections in the region of high frequency as well as the reflector post

of the same marking to be very similar in size and position, we can use the IoU as a good indicator for matching edges.

To match segments and those that are already grouped together, we compute the line that connects the vanishing point from the previous frame and the center of the topmost segment of the marking. Then each unmatched segment is shifted along the x-axis onto this line and the IoU between the shifted and the markings top segment is computed. A segment that belongs to a marking should nicely extend it and, when shifted onto the connection line, should not differ from the topmost segment of the marking and hence have a high IoU.

The process iteratively matches segments bottom to top until all segments are grouped into several marking. In each round, the best match between all unmatched segments and all markings is found and added to a marking if the IoU is at least 0.25. When no such match is found, the bottommost segment forms a new marking. To increase computation speed, the process only considers segments that are not further away than 10 image lines from the topmost segment of the marking. The result of this step is a set of double-edge segments, where each set represent a part of a lane marking.

In Fig. 5, the result of this stage based on the double-edge segments, as mentioned before, can be seen. Segments with the same color belong to the same marking.

3.4 Lane Marking Filtering

After all segments are grouped, each marking is checked against simple constraints and deleted if these are not met. Either the number of segments within the marking is too low or the average width of marking is larger than its height. We found that the minimum number of segments per marking to be 10 is a good value, as well as an aspect ratio threshold of 2. This means every marking that has more than twice its width as height is considered not useful. All remaining markings are used to estimate the vanishing point.

Fig. 5 Results of the double-edge grouping step. Each color represents a group of double-edge segments that likely belong to the same lane marking

In Fig. 6, the results of filtering markings can be seen. Almost all of the false detections on the side are removed due to their small size or overall low intensity.

3.5 Lane Marking Simplification

This section describes the simplification steps performed to reduce the amount of data used in the voting process.

Since, during grouping, it is assumed that line markings are primarily straight, we can reduce the number of points per markings drastically without loss of precision; in an optimal case down to two points. However, since we allowed small deviations from the perfect line, simply keeping the top- and bottommost segments could potentially result in bad approximations. We decided to use a simple scheme to reduce the number of points per markings in a linear way. The new start point of the marking is computed as the average value of the first two original points reprojected to the original image line. The same is done for the last point of the marking. Subsequent support points are computed at every n-th point by averaging all points around the current point within a range of half the distance to the next support point.

From experiments, we have found that a support point distance of 20 lines produces good results, while reducing the number of points to 5% of the full set.

3.6 Vanishing Point Estimation

This section describes how the vanishing point is estimated based on the simplified lane markings.

Each line segment of each marking is extended and two points of intersection are computed. The first one is the intersection of the extended segment with the image row of the previously estimated vanishing point, and the second one is the intersection with the corresponding image column. After all intersection points are

Fig. 6 Results after lane marking filter. It can be seen that most of the false double-edge detections are removed since they do not form shapes after grouping that could be lane markings

computed, the new vanishing point is set to the weighted average of all those intersections. Each intersection is assigned a weight depending on the corresponding marking that caused the intersection.

Marking weights are constructed depending on the markings' appearance and their size. A high intensity as well as a high contrast to the surrounding is a strong indicator for a real marking. Hence, the weight is computed as the product of the average intensity of the double-edge segments and their average gradient value computed as described in Sect. 3.1. Additionally, the weight is multiplied by the length of the marking in order to reduce the weight for short markings that could potentially arise from false detections due to noise in the image.

After the vanishing point for the current frame is computed as described above, it is integrated into the previous estimate using an exponential moving average. This way, the vanishing point is able to adapt to slope changes in the street and is robust against false estimates.

4 Results

We tested our approach on various videos containing city, highway as well as county road scenes. Furthermore, the videos are taken under various weather conditions and daytimes. Some example frames are presented in Fig. 7.

The left side of the images shows the actual scene with the estimated vanishing point position depicted as a red circle, while the right side shows the results of the voting process. Green lines represent the extended lane markings and each gray circle represents an intersection of a marking with either the horizontal or vertical vanishing point line from the previous frame. The green circle represents the weighted average of all intersections, the red circle represents the vanishing point of the previous frame and the blue circle is the integrated vanishing point that is the current estimate.

We have recorded the number of detections, markings and filtered markings as well as computation times to measure the effects of the single stages on the overall runtime. For our experiments, we used videos with a resolution of 1280×720 pixels. The results are averaged over 1000 frames.

The number of double-edge segments that are detected in the first stage is around 1000 segments which are found within circa 5.5 ms per frame. Those segments are then grouped into an average of 230 markings and filtered down to 12 markings within 0.4 ms. Afterward, the marking simplification step takes less than 0.02 ms and the final estimation of the vanishing point takes another 1.5 ms. The overall processing time for a single frame at a resolution of 1280×720 is in total 7.44 ms, which results in a frame rate of more than 140 frames per second.

In comparison with other methods for fast vanishing point estimation, e.g., (Moghadam et al. 2012), our method has various advantages. We are able to

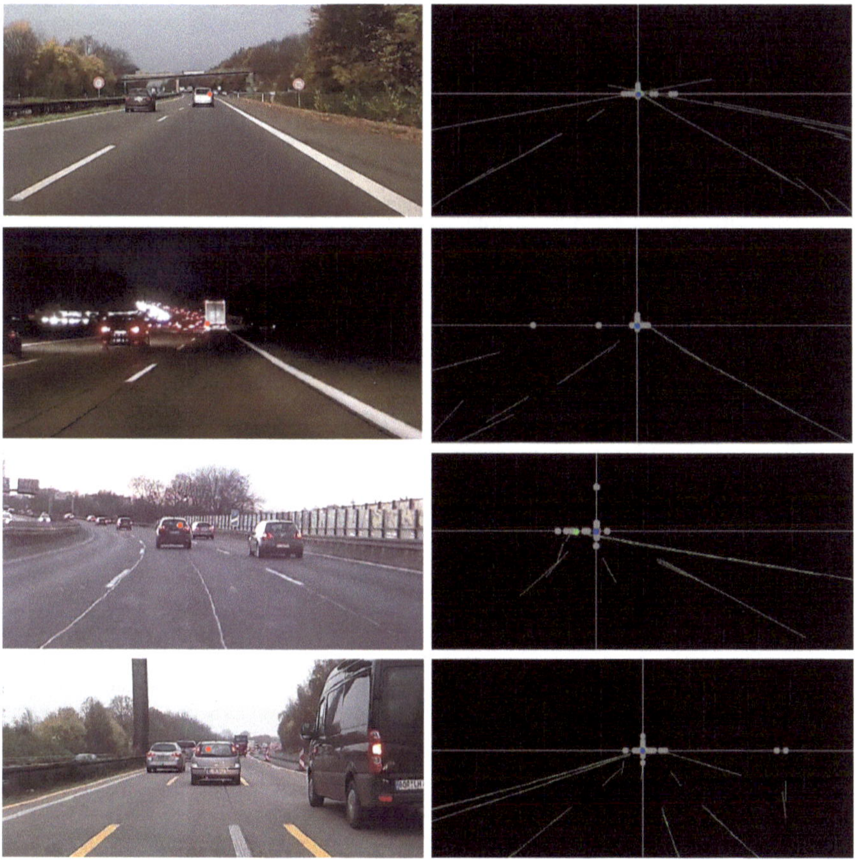

Fig. 7 Example frames from different test videos. It can be seen that the vanishing point is detected reliably across all frames. *Left* Input frames with detected vanishing points as *red circles*. *Right* Vanishing point estimation. *Green lines* represent simplified markings, the *green dot* is the estimated vanishing point, and the *blue dot* is the integrated vanishing point

achieve above real-time performance on a resolution of 1280×720 whereas other algorithms need to down sample drastically, e.g., to 80×60 pixels, to match this requirement. However, the proposed method relies on structured roads, and especially it looks for lane markings, whereas other algorithms can work with arbitrary edge-like structures at the cost of more complex operations, e.g., Gabor filters. This makes the proposed method best applicable to highway and city scenes rather than any natural scene. However, the standard use case is driving cars on structured roads and in general a minimal processing time is favored due to the speed of the vehicles.

5 Conclusion

This paper presented a method to estimate the vanishing point in real time in videos of structured roads. Based on appearance priors, road markings are detected as horizontal line segments limit by a min-max-edge pair that represents the left and right high contrast borders. Next, the detections are grouped based on proximity and similarity measured by the intersection over the union between detections. After grouping, the method filters the resulting markings to remove false detections. To further reduce the amount of data, all markings are approximated by piecewise linear functions. To get to the estimate of the vanishing point, each piece of a marking is extended and its intersection with the image row and column of the previous vanishing point is computed. The vanishing point is then set to the weighted average of all these intersections. The weights represent the quality of the marking, by incorporating its intensity, contrast to the surrounding, and length.

References

Bertozz M, Broggi A, Fascioli A (1998) Stereo inverse perspective mapping: theory and applications. Image Vis Comput 16(8):585–590

Borkar A, Hayes M, Smith M, Pankanti S (2009) A layered approach to robust lane detection at night. In: IEEE workshop on computational intelligence in vehicles and vehicular systems, March 2009, pp 51–57

Deng J (2013) Fast lane detection based on the b-spline fitting. Int J Res Eng Technol:134–137

Lin C, Wang M (2010) Topview transform model for the vehicle parking assistance system. In: International computer symposium, December 2010, pp 306–311

Moghadam P, Starzyk J, Sardha W (2012) Fast vanishing-point detection in unstructured environments. IEEE Trans Image Process 21(1):425–430

Energy-Efficient Driving in Dynamic Environment: Globally Optimal MPC-like Motion Planning Framework

Zlatan Ajanović, Michael Stolz and Martin Horn

Abstract Predictive motion planning is a key for achieving energy-efficient driving, which is one of the major visions of automated driving nowadays. Motion planning is a challenging task, especially in the presence of other dynamic traffic participants. Two main issues have to be addressed. First, for globally optimal driving, the entire trip has to be considered at once. Second, the movement of other traffic participants is usually not known in advance. Both issues lead to increased computational effort. The length of the prediction horizon is usually large and the problem of unknown future movement of other traffic participants usually requires frequent replanning. This work proposes a novel motion planning approach for vehicles operating in dynamic environments. The above-mentioned problems are addressed by splitting the planning into a strategic planning part and situation-dependent replanning part. Strategic planning is done without considering other dynamic participants and is reused later in order to lower the computational effort during replanning phase.

Keywords Eco-driving · Optimal speed trajectory · Dynamic environment · Real-time capability · Replanning

Z. Ajanović (✉) · M. Stolz
Electrics/Electronics and Software, Virtual Vehicle Research Center, Inffeldgasse 21/A, 8010 Graz, Austria
e-mail: zlatan.ajanovic@v2c2.at

M. Stolz
e-mail: michael.stolz@v2c2.at

M. Horn
Institute of Automation and Control, Graz University of Technology, Inffeldgasse 21/B, 8010 Graz, Austria
e-mail: martin.horn@tugraz.at

C. Zachäus et al. (eds.), *Advanced Microsystems for Automotive Applications 2017*, Lecture Notes in Mobility, DOI 10.1007/978-3-319-66972-4_10

1 Introduction

Knowledge of the upcoming driving route, the road conditions, and the ability to control the vehicle's propulsion is an enabler for optimization of the driving behavior with respect to energy consumption. Discrete dynamic programming (DP) has been used for over a decade now for this purpose (e.g., in research focused on heavy-duty vehicles (Hellström 2005, 2010)). A comparison between different optimization methods (Euler–Lagrange, Pontryagin's Maximum Principle, DP, and Direct Multiple Shooting) was presented in Saerens (2012). The work additionally covers an analysis on the DP grid choice, tips on backward and forward dynamic programming, and how to incorporate traffic lights. The authors of Kamal et al. (2011) showed that model predictive control (MPC) enables notable fuel savings for vehicles driving on free roads with up and down slopes. Additional usage of MPC was presented in Vajedi and Azad (2016) for control of a hybrid vehicle driving over a hill and performing vehicle following. An overview of existing approaches treating this as an optimal control problem and the current state of the art can be found in Sciarretta et al. (2015).

The integration of traffic lights into optimal motion planning has also been studied intensively. In Mahler and Vahidi (2012), authors showed an approach for the case of incomplete knowledge about upcoming traffic lights' timing. The case of complete knowledge of the upcoming traffic lights' timing together with Dijkstra's algorithm was studied in Nunzio et al. (2013) and in Kural et al. (2014) an MPC-based controller was developed with additional constraints imposed on a vehicle in front.

The vehicle following problem is studied in Mensing et al. (2013). A possible solution is presented showing different concepts for the safe vehicle following, defining helpful concepts such as the safe distance, time-intervehicular, and time-to-collision. A possible solution for the comfort-oriented vehicle following is presented in Schmied et al. (2016) with leading vehicle movement prediction treated as a disturbance in an MPC controller. Several publications (Wang et al. 2015; Murgovski 2015; Kamal et al. 2016; Shamir 2004) are approaching the planning of optimal overtaking with different goals. With respect to energy efficiency, all these methods modify an optimal speed trajectory in a way so that it leads to the smallest.

The mentioned publications can be roughly grouped into MPC approaches which execute replanning continuously during driving and optimal control approaches which plan the entire trip at once. MPC approaches are generally dealing well with dynamic constraints but cannot guarantee globally optimal solutions as their prediction horizon is limited. On the other hand, optimal control approaches generally guarantee globally optimal solutions for the initial problem, but not in the presence of disturbances. The proposed approach fills the gap between these two approaches using optimal trajectory tree and MPC-like replanning scheme.

2 Problem Definition

This work focuses on an energy-efficient motion planning algorithm based on dynamic programming in the presence of dynamic constraints. Within this, computational efficiency of the algorithm is important for achieving online adaptability. A common approach for achieving energy-efficient driving is to first formulate an appropriate optimal control problem. This problem is then solved offline and the resulting velocity trajectory is used as a reference for low-level speed control.

When applied like this, in real traffic scenarios with other traffic participants or dynamic constraints such as traffic lights, these reference trajectories may not be followed by low-level control. This would directly lead to nonoptimal driving. To avoid this, the motion of other traffic participants has to be considered as a constraint in the optimization problem as it is shown in Ajanovi et al. (2017). Practically, this is not always possible, as the motion of other traffic participants is only known when they are in a sensor field of view of the ego vehicle, but not when the initial planning is done. Additionally, when predicting future motion of other traffic participants, model uncertainties cause deviations between real and predicted driving over time. To avoid deviations, frequent replanning is necessary. This brings significant computational burdens if the whole trip is considered, which is necessary to achieve a globally optimal solution.

2.1 Optimal Control Problem

Formally, this problem can be expressed as an optimal control problem with an appropriate cost function. The cost function has to reflect the aforementioned requirement of minimal energy consumption. This includes energy used for propulsion and energy used onboard (e.g., infotainment, component temperature management, air conditioning). A dynamic vehicle model is used to estimate the propulsion force needed to compensate for resistance forces (gravity, air drag, roll resistance) and to provide the required acceleration. Detailed derivation of the model used in this work is presented in Ajanovi et al. (2017). If only energy used for propulsion is considered, energy-efficient behavior would result in smooth, low-speed driving (almost zero). However, as onboard energy usage is proportional to driving time, slow driving increases the overall consumption. The optimal speed trajectory is, therefore, a balance between these two types of consumption.

In addition, an optimal velocity trajectory has to satisfy several constraints. Constraints can be classified as internal or external. Internal constraints arise from system limitations (e.g., maximum acceleration, velocity, torque) while external constraints are caused by the environment (e.g., traffic signs, other traffic participants). The integration of constraints such as collision avoidance is not straightforward, as these constraints are time- and space varying and depend on the driving trajectory of the controlled vehicle itself.

Although analytical optimization approaches exist, the application to the discussed planning task is problematic, due to incorporating various constraints. The focus, therefore, in the following work is a numeric optimization, especially graph-based approaches, since they are the most flexible and applicable to the nonlinear problem.

2.2 Computational Complexity

To solve the optimization task numerically, using graph searching methods, state discretization is necessary. By increasing the number of system states considered in the optimization problem, complexity is increased exponentially, as the number of possible state combinations increases exponentially. Each additional state multiplies the number of state combinations by the number of its discretization levels. Additionally, and even more problematic, the number of possible transitions needed to be evaluated in each step is increased significantly. Bellman called this problem the "*curse of dimensionality*".

Other traffic participants represent time and space-varying constraints on both position and velocity. Because of that, travel time must be used as a system state, besides travel distance and velocity. All three system states must be discretized and the whole state space must be searched to achieve a globally optimal solution. This requires significant computational effort.

3 Optimal Motion Planner

The main idea of the optimal motion planner introduced in this work is based on the combination of the advantages of forward and backward dynamic programming (Fig. 1). The planning problem is addressed by splitting into strategic planning and situation-dependent replanning. The results once calculated by backward programming, in strategic planning phase, are continuously reused for the ongoing replanning during driving. Replanning is done using forward planning from the actual system state, for a certain prediction horizon into the future, and merged with previously obtained results from backward planning. During replanning, dynamic constraints and additional states (e.g., lanes, travel time) are considered. In this way, the whole trip is taken into consideration along with newly arisen constraints, but only a planning for a defined horizon is executed. This promises benefits of both forward planning the entire trip (globally optimal solution) and adaptability of MPC with a significant reduction in computational effort.

3.1 Dynamic Programming

Dynamic programming is a preferred method used for solving the optimal control problem discussed in this work. The main advantages are its flexibility and possibility to incorporate different kinds of models and constraints and the fact that it results in a globally optimal solution. It is based on the Principle of Optimality, introduced by R. Bellman (Ajanovi et al. 2017).

The iterative approach of dynamic programming can be executed starting from the goal state toward the initial state (backward dynamic programming) and vice versa (forward dynamic programming). The advantage of the backward calculation is that the calculated result can be reused during the trip, as it only depends on the final state. This is not the case with the forward calculation, where results are related to specific initial states. On the other hand, the advantage of forward calculation is that other states such as the time of travel can be calculated as the initial time is always known.

3.2 Strategic Planning

The strategic planning phase is executed only once at the beginning of a trip or if the target location changes. It is achieved using backward dynamic programming starting from a goal state, backward in space. In this phase, only time-invariant constraints are considered (e.g., speed limits) with topological road profile and vehicle model. The results of this phase are *the optimal trajectory tree*, *the cost-to-go map,* and *the initial optimal velocity trajectory*. The initial optimal velocity trajectory is one branch of the optimal trajectory tree which passes through the initial state.

Optimal speed trajectory tree

The optimal speed trajectory tree is a tree-like structure formed by connecting all optimal transitions by lines. Together with a cost-to-go map, it gives insight into the optimal behavior when only static constraints are present.

It can be noted generally that if two different trajectories have a common node they will continue on the same trajectory toward the goal. This implies that when planning a trajectory in forward approach if constraints introduced by other traffic participants are not active anymore, a trajectory from a backward planning starting from that state toward the goal can be reused. In this work, we will use this property of the optimal trajectory tree to reduce the computational effort needed.

An optimal speed trajectory tree for a problem considered in this work with discretization steps of 5 m for distance and 0.5 m/s for speed is shown in Fig. 2. This map is generated from the goal state toward the start using backward DP. Additionally to the initial optimal trajectory for the given initial condition, multiple other trajectories (branches) for different initial conditions are available.

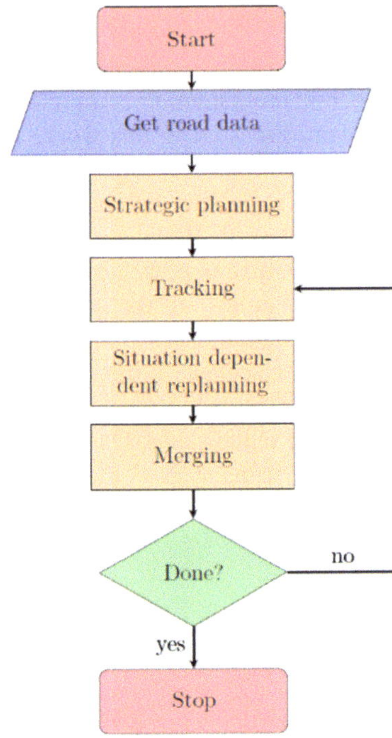

Fig. 1 Optimal motion planner flowchart

Cost-to-go map
The cost-to-go map provides additional information to the optimal trajectory tree. It represents the minimum energy needed to finish a trip from that state point. It can be achieved by following an optimal trajectory, represented as a branch on the optimal trajectory tree starting from that state point. In Fig. 3 the cost-to-go map for the same problem as in Fig. 2 is shown.

3.3 Situation-Dependent Replanning

During the replanning phase, the optimal trajectory is adjusted by taking into consideration dynamically arisen constraints. The adjustment is done by replanning the optimal trajectory in an efficient way by reusing the cost-to-go map and an optimal speed trajectory tree. The replanning is done with forward dynamic programming starting from the actual system state in operational space for a defined prediction horizon in the future.

Several safety factors such as maximum time of overtaking execution (constraint on minimum velocity difference), minimum distance from the leading vehicle and clearance needed for lane changing are considered as constraints in this phase.

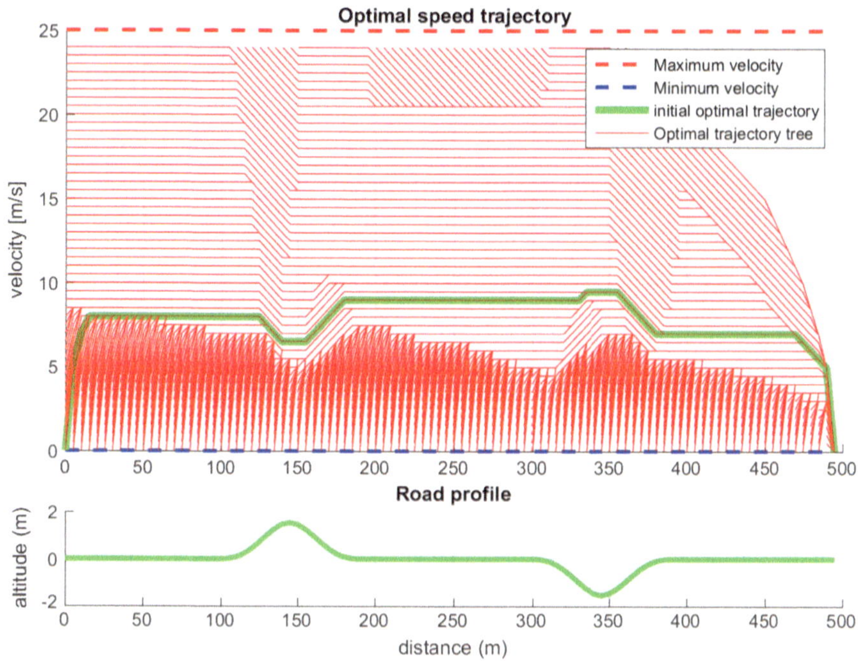

Fig. 2 Optimal speed trajectory tree

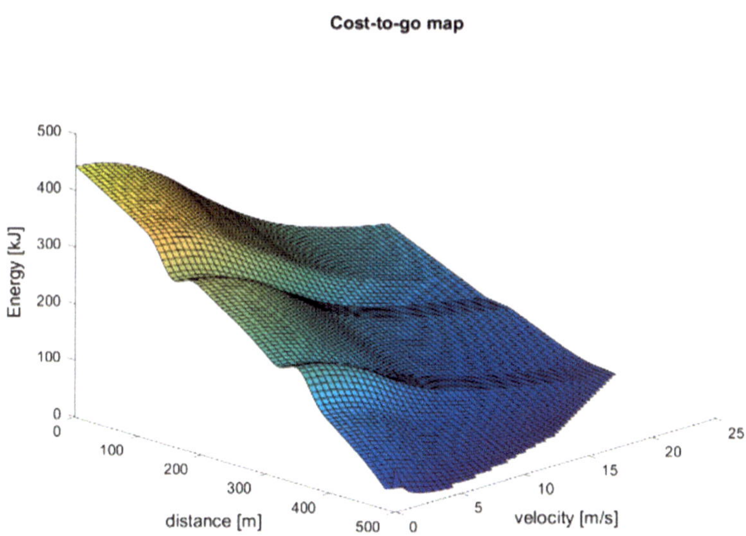

Fig. 3 Cost-to-go map

The future movement of other vehicles is calculated using a simple prediction model that assumes that the leading vehicle will continue moving with constant speed and that it will slow down if it reaches the controlled vehicle (after being overtaken). More sophisticated models of the leading vehicle's velocity which may depend on space, time, and the controlled vehicle can be also included.

A principle of operation is shown in Fig. 4. Gray lines represent the optimal trajectory tree, constructed in the strategic planning phase. The blue line is the initial optimal trajectory, which also results from the strategic planning phase. The vehicle drives on the initial optimal trajectory until situation-dependent replanning is initiated. Using forward planning, a forward optimal trajectory tree starting from the actual state is constructed (dashed lines) considering dynamic constraints such as other traffic participants, traffic lights, etc.

The merging of two trajectory trees (forward and backward) is done at the end of the forward replanning phase. *Cost-to-come*, the *cost-to-go* equivalent in forward planning, values at the *possible joining nodes* at the end of the replanning horizon are summed with *cost-to-go* cost at these nodes from backward planning. In this way, combined costs of moving on trajectories partially planned forward and partially planned backward are calculated. The minimum among these costs is chosen, defining an *optimal joining node*, on the new optimal trajectory (solid black node). Starting from this node backward, toward the actual state, a *new optimal trajectory* can be constructed iteratively. A *new optimal trajectory* does not have to be constructed fully to the final state if the new replanning happens while driving within the prediction horizon. In this way, unnecessary calculations can be avoided.

In Fig. 4, no additional system states (e.g., time, lane) are drawn within forward planning to keep clarity of working principle. Eventually, additional states can be

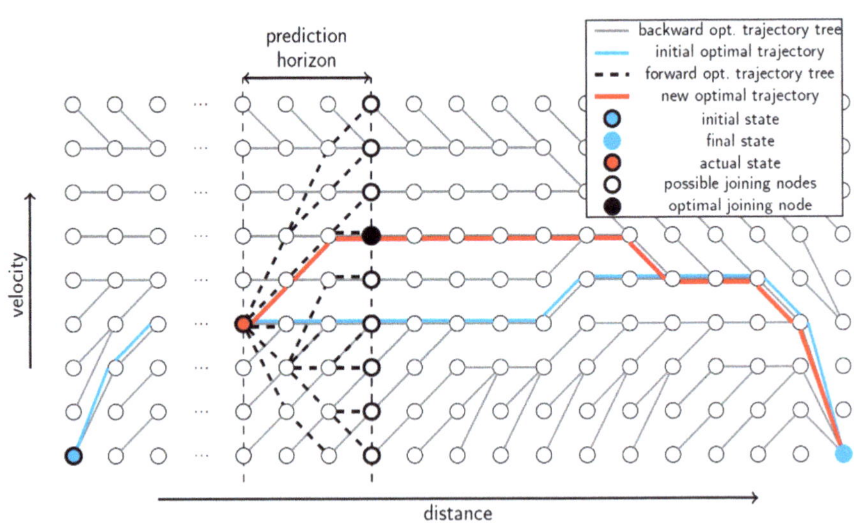

Fig. 4 Situation-dependent replanning

visualized on the third dimension in parallel to presented forward planning. During merging, for choosing the optimal joining node, the values of these states are neglected and an appropriate node from cost-to-go is chosen based on the velocity and position state values only.

The situation-dependent replanning procedure is repeated during the entire trip. It can be triggered by either spatial length traveled, time period, by an event (e.g., detecting of other traffic participants, detecting a significant deviation of predicted motion of other traffic participants) or a combination of these.

3.3.1 Prediction Horizon

As spatial discretization is used as a basis for motion planning, the prediction horizon is defined by the length at which a forward planning is executed. It is important to choose a proper horizon length as a so-called short sightedness can appear otherwise. Several constraints such as clearance needed for changing a lane can block lane changing if horizon length is not enough. In general, the horizon should be as long as computational resources enable it, but keeping in mind that accuracy of motion prediction of other traffic participants decreases with time.

Horizon length depends also on the replanning length as it should be long enough, so that, when next replanning is initiated, enough clearance for lane change is guaranteed (if it exists).

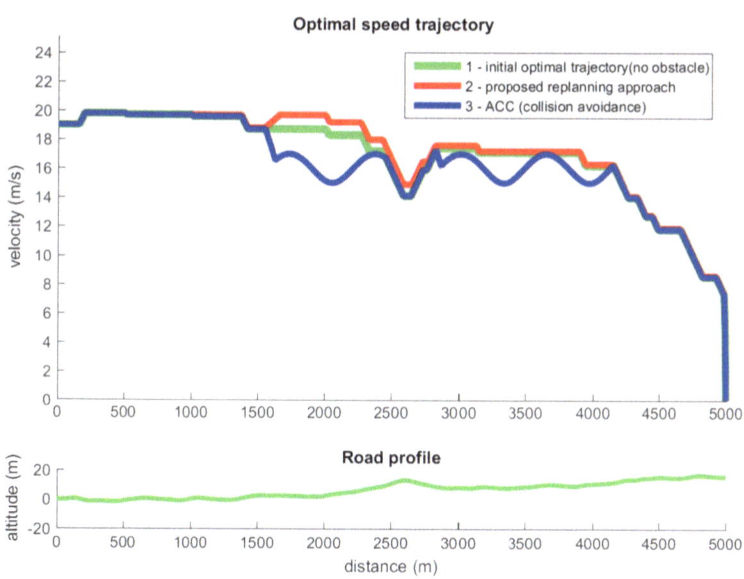

Fig. 5 Simulation results

3.3.2 Replanning Triggering

The replanning triggering is most likely be determined by a length. This means that the replanning frequency is not constant, as it depends on driving velocity. For real-life application, this is not a big issue for driving on the highway, as there is no big fluctuation of velocity. Frequent replanning is important if the environment is highly dynamic and the prediction of other traffic participant's motion is not precise. Generally, it is better if replanning length is shorter (replanning frequency higher), but this increases computational effort. Finally, a trade-off of these two has to be made.

4 Simulation Results

To present advantages of this contribution, a realistic driving scenario is simulated (Fig. 5). A vehicle is driven on an optimal trajectory on a multilane road and approaches a slower moving vehicle, with an average velocity of 16 m/s with sinusoidal fluctuations of amplitude 1 m/s and period 40 s. This unmodeled acceleration introduces a deviation from the predicted motion, as constant velocity model (CVM) is used to predict future motion. The measured actual velocity of other traffic participants, at the time of replanning, is used for prediction during planning. In forward planning, the ego vehicle has to drive at least 3 m/s faster to overtake other vehicles.

Because of the deviation, the planned trajectory can lead to a collision. Therefore, the ego vehicle is equipped with an ACC, so if the desired trajectory would result in a collision, the ACC would slow down the vehicle. Unfortunately, this causes additional energy consumption as the vehicle is deviating from its initial optimal speed trajectory. Therefore, frequent replanning is necessary to adjust the optimal trajectory to new situations and achieve optimal driving.

To analyze the potential for energy savings, different motion planning approaches are simulated and compared. The results are shown in Table 1. The first approach was initial optimal trajectory without considering obstacles. The second approach is the approach introduced in this work. The third approach is using only initial optimal trajectory and relies on ACC to adapt velocity to avoid collisions.

Table 1 Different motion planning approaches

#	Replanning length [m]	Horizon length [m]	Driving time [s]	Energy consumed [kJ]	Energy cons [%]
1	–	–	309,77	2673,1	0
2	25	250	305,05	2679,9	+0.25
3	–	–	320,57	2842,3	+6,32

5 Conclusion

The example reveals that eco-driving algorithms can provide significant energy savings. To be accepted by drivers and to achieve real driving benefits, eco-driving algorithms should be intuitive and adaptable to dynamic environments. They should rely only on the information available onboard at the time of planning. The simulation example showed that using only the initial optimal trajectory planned at the start of a trip is far from optimal solution in dynamic environments.

The presented approach overcomes these problems and presents a novel approach for optimal motion planning in dynamic environments. Reusing cost-to-go map and optimal trajectory tree effectively provides complete trip planning benefits.

It was shown that the proposed approach allows continuous adjustment to dynamic environments, when information is unknown at initial planning. Additionally, problems arising from deviation in the prediction of other traffic participant movement are compensated by frequent replanning. Forward planning for the whole trip leads to lowest energy consumption, but this approach is unrealistic as complete information is usually not available, and calculation time is significantly longer. Energy consumption in the proposed approach is only negligibly higher.

The novel approach proposed in this work promises an intuitive and adaptable solution for real-life eco-driving.

Acknowledgements The project leading to this study has received funding from the European Union's Horizon 2020 research and innovation program under the Marie Skłodowska-Curie grant agreement No 675999, ITEAM project.

VIRTUAL VEHICLE Research Center is funded within the COMET—Competence Centers for Excellent Technologies—program by the Austrian Federal Ministry for Transport, Innovation and Technology (BMVIT), the Federal Ministry of Science, Research and Economy (BMWFW), the Austrian Research Promotion Agency (FFG), and the province of Styria and the Styrian Business Promotion Agency (SFG). The COMET program is administrated by FFG.

References

Ajanović Z, Stolz M, Horn M (2017) Energy efficient driving in dynamic environment: considering other traffic participants and overtaking possibility. In: Comprehensive energy management—eco routing and velocity profiles. Springer International Publishing, pp 61–80

Hellström E (2005) Explicit use of road topography for model predictive cruise control in heavy trucks. MS thesis, Linköping University, Sweden

Hellström E (2010) Look-ahead control of heavy vehicles, PhD thesis, Linköping University, Linköping, Sweden

Kamal MAS, Mukai M, Murata J, Kawabe T (2011) Ecological vehicle control on roads with up-down slopes. IEEE Trans Intell Transp Syst 12(3):783–794

Kamal MAS, Taguchi S, Yoshimura T (2016) Efficient vehicle driving on multilane roads using model predictive control under a connected vehicle environment. IEEE Trans Intell Transp Syst PP(99):1–11

Kural E, Jones S, Parrilla AF, Grauers A (2014) Traffic light assistant system for optimized energy consumption in an electric vehicle. In: International conference on connected vehicles and expo (ICCVE)

Mahler G, Vahidi A (2012) Reducing idling at red lights based on probabilistic prediction of traffic signal timings. In: 2012 american control conference (ACC), Montreal

Mensing F, Bideaux E, Trigu R, Tattegrain H (2013) Trajectory optimization for eco-driving taking into account traffic constraints. Transp Res Part D Transp Environ 18:55–61

Murgovski JSN (2015) Predictive cruise control with autonomous overtaking. In: 54th IEEE conference on decision and control (CDC), Osaka

Nunzio GD, Wit CCD, Moulin P, Domenico DD (2013) Eco-driving in urban traffic networks using traffic signal information. In: 52nd IEEE conference on decision and control, Florence, Italy, 10–13 December 2013

Saerens B (2012) Optimal control based eco-driving. PhD thesis, Katholieke Universiteit Leuven, Belgium

Schmied R, Waschl H, Re LD (2016) Comfort oriented robust adaptive cruise control in multi-lane traffic conditions. In: 8th IFAC international symposium on advances in automotive control, Norrköping, Sweden

Sciarretta A, Nunzio GD, Ojeda L (2015) Optimal ecodriving control: energy-efficient driving of road vehicles as an optimal control problem. IEEE Control Syst Mag:71–90

Shamir T (2004) How should an autonomous vehicle overtake a slower moving vehicle: design and analysis of an optimal trajectory. IEEE Trans Autom Control 49:607–610

Vajedi M, Azad NL (2016) Ecological adaptive cruise controller for plug-in hybrid electric vehicles using nonlinear model predictive control. IEEE Trans Intell Transp Syst 17(1):113–122

Wang M, Hoogendoorn S, Daamen W, Arem BV, Happee R (2015) Game theoretic approach for predictive lane-changing and car-following control. Transp Res Part C Emerg Technol 58(Part A): 73–92

Part III
Data, Clouds and Machine learning

Automated Data Generation for Training of Neural Networks by Recombining Previously Labeled Images

Peter-Nicholas Gronerth, Benjamin Hahn and Lutz Eckstein

Abstract In this paper, we present our approach to data generation for the training of neural networks in order to achieve semantic segmentation in an autonomous environment. Using a small set of previously labeled images, this approach allows to automatically increase the amount of training data available. This is achieved by recombining parts of the images, while keeping the overall structure of the scene intact. Doing so allows for early network training, even with only few training samples at hand. Furthermore, first results show that training networks using the so created datasets allow for good segmentation results when compared to publicly available datasets.

Keywords ADAS · Semantic segmentation · Convolutional neural networks · Semi artificial datasets · Training data generation · Image stitching

1 Introduction

With the current move toward the development of automated driving a growing number of vehicle sensors can be observed in order to detect the environment around the vehicle as well as other traffic participants. In the multitude of different sensors available, cameras have become one of the standard options, as they can be used in a variety of applications while being available at comparably low prices.

P.-N. Gronerth (✉) · L. Eckstein
Institut für Kraftfahrzeuge, RWTH Aachen, Steinbachstraße 7, 52074 Aachen, Germany
e-mail: peter.gronerth@ika.rwth-aachen.de

L. Eckstein
e-mail: office@ika.rwth-aachen.de

B. Hahn
fka Forschungsgesellschaft Kraftfahrwesen mbH Aachen, Steinbachstraße 7, 52074 Aachen, Germany
e-mail: benjamin.hahn@fka.de

C. Zachäus et al. (eds.), *Advanced Microsystems for Automotive Applications 2017*,
Lecture Notes in Mobility, DOI 10.1007/978-3-319-66972-4_11

In most image processing approaches the camera image is first processed, aiming toward the detection of different elements. Afterwards, additional information (e.g., lanes and their course, traffic lights and their status or other road users lie cars, trucks or pedestrians and their speed, position and type) can be generated. One potential method to create such detections in an image is via segmentation.

During this process, every pixel in an image gets assigned to a predefined class, representing the type of object it belongs to. For an example of such image combinations see Fig. 1 above.

In recent years, image segmentation has become a standard application for convolutional neural networks. To do so, fully segmented images representing the ground truth need to be created. This is done mostly manually, with a very high pixel precision. These images are then used as desired output during training, with the aim to "teach" a network to generate such information on its own. Due to the potentially high number of labeled images necessary to train the network, this can easily become a very time- and labor-consuming process, of up to 90 min per picture (Brostow et al. 2009; Cordts et al. 2016).

We propose to use a recombination of already labeled images, to generate new automatically labeled images. Applied, for example, during early stages of development, this can be used to heavily increase the amount of data available for training. Furthermore, the approach presented in this paper, is heavily automated to minimize the amount of time necessary to achieve the increase in data available for training. Comparing networks with a fixed structure trained over different publicly available real and fully artificial datasets shows that the approach in this paper leads to segmentation results of a better quality.

The paper is structured as follows. In the second chapter, references to related work and the state of the art can be found, along with a brief overview on datasets considered for the comparison during the evaluation. The third chapter describes the overall process and routines used to create our semi-realistic dataset. In the fourth chapter, a short comparison between different training sets is performed. In the final chapter, the results are summarized and future development steps are presented.

Fig. 1 Camera image and exemplary fully segmented counterpart

2 Related Work

As this paper mostly concentrates on the detailed process of the dataset creation, a detailed overview and introduction regarding (convolutional) neural networks and learning image segmentation is omitted, but can be taken from the following sources (Szegedy et al. 2015; Arbeláez et al. 2012; Carreira et al. 2012; Krizhevsky et al. 2012; Girshik et al. 2014; Long et al. 2015).

2.1 *Available Public Datasets*

As neural networks have become more and more of interest, multiple datasets have been made publicly available for testing and evaluating new structures or training methods. In this chapter, a brief overview on five datasets, focused on an automotive application, is given. A corresponding list can be taken from Table 1.

The KITTI dataset was developed at the Karlsruhe Institute for Technology and the Toyota Technological Institute in Chicago. It contains 7 481 training and 7 518 test images with a total of 80 256 objects, categorized into 9 different classes (Geiger et al. 2013). The objects are noted as bounding boxes, a segmented representation of the images is only available from third parties.

A second publicly available dataset is called CamVid (Cambridge-driving Labeled Video Database) and is comprised of 701 images. They are labeled by hand and each pixel belongs to one of 32 different segmentation classes. The resolution is at 960 × 720 pixels (Brostow et al. 2009).

The Cityscape dataset, provided by TU Darmstadt, Max Plank Institute for Informatics and the Daimler AG R&D group, consists of street scenes collected in 50 different cities. It consists of two groups a densely labeled set of 5 000 images and a larger set of 20 000 weakly annotated frames. Both contain up to 30 different segmentation classes. It furthermore provides additional images for stereo vision, GPS coordinates, ego-motion data, and outside temperature. Please refer to Cordts et al. (2016) for more information regarding this dataset.

In contrast to these three datasets, which are all fully comprised of real camera images, the SYNTHIA (synthetic collection of imagery and annotations) dataset only contains fully artificial images. It consists of more than 200 000 images divided into multiple light and seasonal conditions and scenarios. In total, 13

Table 1 Considered datasets

Name	Type	Selected	Dataset size
KITTI	Real images	No	7 481 (training) + 7 518(test)
CamVid	Real images	Yes	701 (used)
Cityscape	Real Images	Yes	5 000 (used) + 20 000
SYNTHIA	Artificial	Yes	200 000 (used 5 364)
PlayForData	Artificial	Yes	24 966 (used)

different segmentation classes are provided along with object instantiation, distance information, and ego vehicle motion (Ros et al. 2016).

Another fully artificial collection of training data is provided by the TU Darmstadt and the Intel Labs. In the corresponding paper (Richter et al. 2016) the process of extracting images from a computer game is described. The images have a resolution of 1 914 × 1 052 pixels and are labeled using the same classes as in the Cityscape dataset.

2.2 *Image Manipulation and Recombination*

The process of recombining images, has been well developed in the past years and found a broad application in today's world. The key concepts are used when creating panorama images from single frames, combining multiple photographs taken from a group of people to create a single image with a visually preferable look, or to remove parts of images that for example contain unwanted elements. Most systems therefore aim to generate images that are as real as possible.

One such approach is presented by Cho et al. (2008), (2010). It works by subdividing the image into subparts and filling the gaps only with "patches" that are in accordance with an underlying Markov-Network. The results show that an automated approach to image manipulation by removing or adding parts and completely changing the position of given elements of the image is possible. By optimizing the boundary areas, where the original image and the applied manipulation overlap, the quality of the images can be improved further as shown by Tao et al. in (2010).

A different approach, relying on depth information is presented by Mansfield et al., which allows for an object removal or replacement while keeping the overall consistency of the scene intact (Mansfield et al. 2010). An alternative usage of such information is shown by Iizuka et al. in Iizuka et al. (2014), which allows a replacement with the overall perspective in mind. Among other features, the developed system shrinks or magnifies objects in accordance to their new position and manipulates object shadows in accordance with the structure of the new area they overlap.

3 Semi-artificial Dataset Creation

In this chapter, our approach on the creation of a dataset aiming towards training neural networks in semantic segmentation, is described in detail. Furthermore, some exemplary images are presented.

Aiming toward the easy increase in the current amount of data, some images need to be labeled beforehand. This is called "base-set". Using our own labeling tool, specially designed for road scenes, it takes a well-trained labeler roughly 25 min to label an image. The so-created (labeled) images can be categorized into two different types. The first type are images preferably with a high density of fully visible dynamic objects (all traffic participants), called *donor-images*. The second type contains images of road stretches, which do not contain any, or only few and distant traffic participants, called *background-images*. *Donor-images* need to contain information on the position of the lane markings as well as either bounding boxes around or a segmented representation of the dynamic objects. Using this, we take the images apart and extract every single dynamic object into its own image. Parallel to each such sub-image we save a potentially optimized mask to make the removal of the old background as easy as possible and add all necessary information to an access database (see Table 2 for more information and Fig. 2 for a visualization of the sub-image and image mask).

In contrast to the potentially only sparsely labeled *donor-images*, the *background-images* need to be fully labeled with all potential classes that should be detectable by the trained network.

The next step is the actual database creation. We use an xml-styled scene description containing the numbers of dynamic objects to be added, the desired object type and the minimal and maximal distance at which such an object could be placed for every lane in a given *background-image*. We then pick at random the desired number of cars or trucks originally placed at a similar lane in the *donor-image* from our database and warp the sub-image accordingly to a new, potentially random position. These sub-images are then copied in order, from furthest to closest position, to the current *background-image* and the corresponding label file.

This process is repeated as often as necessary, to reach a user specified dataset size. Afterwards the dataset consisting of the semi-artificial images and the ground truth labels can be used during training. Examples for images and labels created this way can be found in Fig. 3 below.

Table 2 Database entry

Name	Data type	Description
ID	Integer	Object identifier
Class	Text	Class of the object
Left, bottom, right, top	Integer	Bounding box position in the donor-image
Distance	Double	Distance estimation between camera and object
Camera	Text	Name of the camera used to acquire the image
Path	Text	Path to the mask and sub-image files

Fig. 2 Exemplary sub-image, image-mask, and combination

Fig. 3 Semi-artificial images created using the method presented in this paper

4 Evaluation

To get a first estimate on the training quality, when using the approach presented in this paper, different publicly available datasets have also been used to train neural networks (an overview can be taken from Table 1).

The KITTI dataset is currently omitted from the evaluation, as labeled images are only available from third parties. This leaves a total of four datasets to compare against, namely SYNTHIA, PlayForData, CamVid, and Cityscape.

The SYNTHIA dataset has been split and only the highway packages have been used (SEQS-01-FALL, SEQS-01-SPRING, SEQS-01-SUMMER, SEQS-06-SPRING, and SEQS-06-SUMMER). The CamVid and PlayForData datasets have been used with all available data. From the cityscape dataset, only the 5 000 densely labeled images were used. Exemplary images from those 4 datasets can be found in Fig. 4 below.

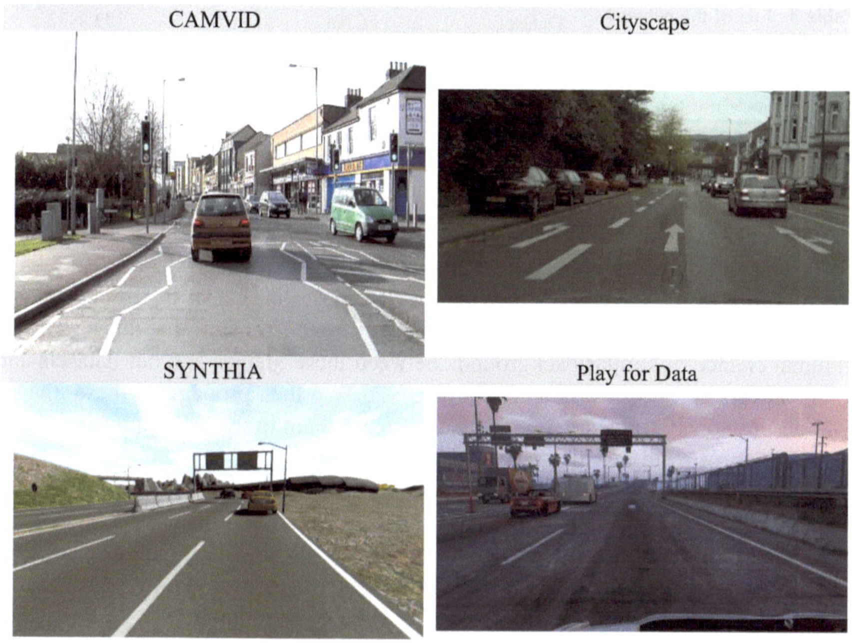

Fig. 4 Exemplary training images from the datasets used during evaluation

We use the process described in the previous chapter to create three different datasets with completely different sub-images and *background-images* from alternative stretches of highway. The first set was created using 48 *background-images* and 852 sub-images and is used as training set with a total of 1 520 images. The second set is created using 12 *background-images* and 212 sub-images to generate a total of 380 images for evaluation during training. The third and final set is created using 5 *background-images* and 50 sub-images and used as test set after training with a total of 100 images. All *background-images* were collected with the same camera, but on different days and at a slightly different camera placement compared to the *donor-images*.

All five datasets are used to train convolutional networks following the FNC8, FCN16, and FCN32 architecture, as proposed by Long et al. in Long et al. (2015). We use an 80% training and 20% validation split and, prior to training, remove 100 images for evaluation from each dataset. The network training itself is carried out using Nvidia Digits utilizing a Titan X and based on the pre-trained Caffe models provided by Long et al. via GitHub. Furthermore, we automatically change the labeled representation of all datasets to an identical definition and resize all images to the same resolution. The aspect ratio is kept by adding black bars to the image boundaries if necessary, which are ignored by the network during training.

After training we generate the *intersection over union (IoU)* for all networks over the 100 evaluation images taken from the corresponding dataset. By doing so, we ensure that the training worked as planned and to get a first look at the overall segmentation quality. The results of this can be taken from Table 3 below.

Table 3 IoU of the networks over the evaluation images

Dataset	FCN8	FCN16	FCN32
CamVid	89.8	86.6	83.1
Cityscape	87.2	85.8	83.9
SYNTHIA	80.5	80.5	77.5
PlayForData	82.5	93.3	92.1
Our dataset	88.0	84.6	82.6

To generate the first evaluation results, 100 new images are labeled. They were recorded the same day and at the same camera position, but at different highway locations compared to the *donor-images*. This is done to ensure that there is only a minimal chance of similar backgrounds between these images and our datasets and no direct repetition of trucks and cars observed. We then propagated these images over all previously trained networks and compare them to the created ground truth. A collection of resulting images can be taken from Fig. 5.

The overall IoU (see Table 4) shows that the realistic datasets perform surprisingly well on our evaluation data, especially when taking into consideration, that they differ greatly in their background. It can also be stated that the artificial datasets perform considerably worse, but further investigation regarding a mixture of artificial and realistic datasets still needs to be carried out.

Our own semi-artificial dataset shows that our approach leads to a well-adapted network, while also reducing the amount of labeling work necessary. The comparably good results can be explained by the greater similarity between the images, due to the usage of the same camera and a similar camera placement between the evaluation and training data.

In addition to the similarities regarding image acquisition, our dataset contains a better representation of the highway scenario, as can be seen by the values reported in Table 5. Since the SYNTHIA dataset has no truck labels and these are only seldom observed in the CamVid and Cityscape scenarios, our dataset outperforms every other dataset in this area. Regarding the car label, it can be observed, that the Cityscape dataset slightly outperforms our training data in case of FCN8 and FCN16 networks, most likely due to the potential higher diversity of cars in this dataset.

5 Summary and Outlook

In this paper, we presented an approach on the creation of semi-artificial datasets. By reusing *donor-images*, previously labeled with either bounding boxes or via semantic segmentation, as well as newly labeled *background-images,* we automatically generate a new dataset.

The presented approach uses a scene definition file and knowledge on the perspective gained by the segmentation of the *background-images*. We furthermore show that networks trained with our datasets, perform well for segmentation tasks of images, taken with an identical camera and at a similar camera position. We

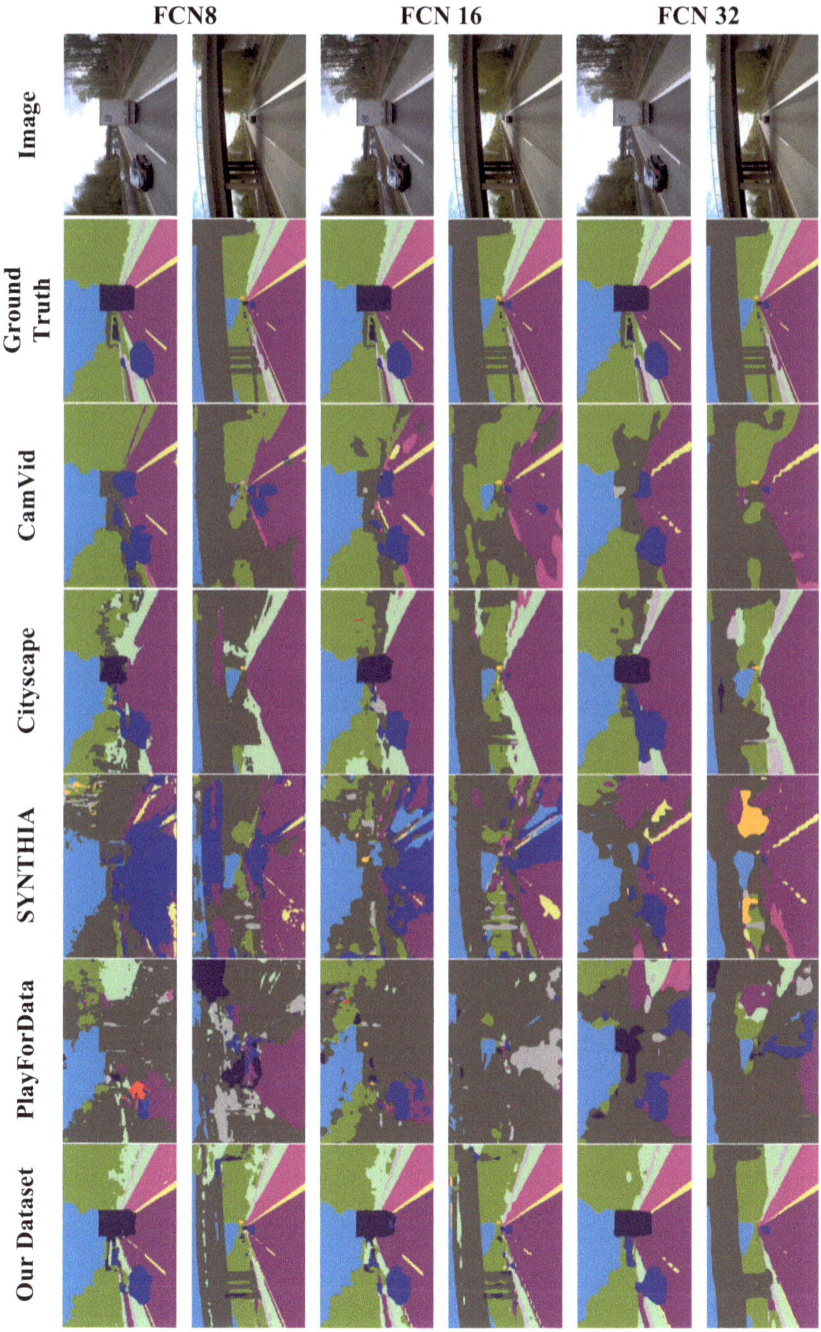

Fig. 5 Segmentation result using the different datasets

Table 4 IoU of the networks over the newly labeled images

Dataset	FCN8	FCN16	FCN32
CamVid	42.0	53.9	50.1
Cityscape	51.3	57.4	63.0
PlayForData	13.5	15.6	21.4
SYNTHIA	37.9	38.5	38.6
Our dataset	81.9	80.9	79.9

Table 5 IoU of the networks regarding the truck/bus and vehicle class over the 100 newly labeled images

Datasddddet	FCN8		FCN16		FCN32	
	Truck	Car	Truck	Car	Truck	Car
CamVid	3.1	22.0	3.8	24.3	0.1	53.8
Cityscape	16.1	57.6	13.3	51.1	28.8	54.8
PlayForData	9.1	21.6	10.1	23.4	6.6	14.3
SYNTHIA	0.0	7.1	0.0	8.4	0.0	26.1
Our dataset	63.0	56.4	63.0	47.2	62.5	56.7

furthermore showed that these networks can outperform networks trained with alternative publicly available datasets, due to the greater similarities between image acquisition and scenarios at hand.

Our next steps in development will lead toward a deeper analysis of the results presented in this paper and a comparison of dataset combinations. Furthermore, we are going to optimize the image generating process, by implementing some of the algorithms presented in the related work chapter. Finally, another set of *background-images* from an urban scenario will be created, aiming toward a better comparison of our results with the existing public datasets.

References

Arbeláez P, Hariharan B, Gu C, Gupta S, Bourdev L, Malik J (2012) Semantic segmentation using regions and parts. In: IEEE conference on computer vision and pattern recognition (CVPR)

Brostow G, Fauquer J, Cipolla R (2009) Semantic object classes in video: a high-definition ground truth database. Pattern Recogn Lett 30:88–97

Carreira J, Caseiro R, Batista J, Sminchisescu C (2012) Semantic segmentation with second-order pooling. In: Computer vision – ECCV 2012. Lecture Notes in Computer Science, vol 7578, pp 430–443

Cho T, Butman M, Avidan S, Freeman W (2008) The patch transform and its applications to image editing. In: IEEE conference on computer vision and pattern recognition

Cho T, Avidan S, Freeman W (2010) The patch transform. IEEE Trans Pattern Anal Mach Intell 32:1489–1501

Cordts M, Omran M, Ramos S, Rehfeld T, Enzweiler M, Benson R, Franke U, Roth S, Schiele B (2016) The cityscapes dataset for semantic urban scene understanding. In: IEEE conference on computer vision and pattern recognition (CVPR), pp 3213–3223

Geiger A, Lenz P, Stiller C, Urtasun R (2013) Vision meets robotics: the KITTI dataset. Int J Robot Res 32(11):1231–1237

Girshik R, Donahue J, Darrell T, Malik J (2014) Rich feature hierarchies for accurate object detection and semantic segmentation. In: IEEE conference on computer vision and pattern recognition

Iizuka S, Endo Y, Hirose M, Kanamori Y, Mitani J, Fukui Y (2014) Object repositioning based on the perspective in a single image. Comput Gr Forum 33(8):157–166

Krizhevsky A, Sutskever I, Hinton G (2012) Imagenet classification with deep convolutional neural networks. Adv Neural Inf Process Syst 25:1097–1105

Long J, Shelhamer E, Darrell T (2015) Fully convolutional networks for semantic segmentation. In: IEEE conference on computer vision and pattern recognition (CVPR), pp 3431–3440

Mansfield A, Gehler P, Gool L, Rother C (2010) Scene carving: scene consistent image retargeting. Lecture Notes in Computer Science, vol. 6311, pp 143–156

Richter S, Vineet V, Roth S, Koltun V (2016) Playing for data: ground truth from computer games. In: European conference on computer vision (ECCV), pp 102–118

Ros G, Sellart L, Materzynska J, Vazquez D, Lopez AM (2016) The SYNTHIA dataset: a large collection of synthetic images for semantic segmentation of urban scenes. In: IEEE conference on computer vision and pattern recognition (CVPR), pp 3234–3243

Szegedy C, Liu W, Jia Y, Sermanet P, Reed S, Anguelov D, Erhan D, Vanhoucke V, Rabinovich A (2015) Going deeper with convolutions. In: IEEE conference on computer vision and pattern recognition

Tao M, Johnson M, Paris S (2010) Error-tolerant image compositing. Lecture Notes in Computer Science, vol 6311, pp 31–44

Secure Wireless Automotive Software Updates Using Blockchains: A Proof of Concept

Marco Steger, Ali Dorri, Salil S. Kanhere, Kay Römer, Raja Jurdak and Michael Karner

Abstract Future smart vehicles will employ automotive over-the-air updates to update the soft ware in the embedded electronic control units. The update process can affect the safety of the involved users, thus requires a comprehensive and elaborate security architecture ensuring the confidentiality and the integrity of the exchanged data, as well as protecting the privacy of the involved users. In this paper, we propose an automotive security architecture employing Blockchain to tackle the implicated security and privacy challenges. We describe our proof-of-concept implementation of a Blockchain-based software update system, use it to show the applicability of our architecture for automotive systems, and evaluate different aspects of our architecture.

Keywords Automotive security architecture · Blockchains · Wireless software update · Over-the-air updates · Security · Privacy · Scalability

---

M. Steger (✉) · M. Karner
Virtual Vehicle Research Center, Inffeldgasse 21/a, 8010 Graz, Austria
e-mail: marco.steger@v2c2.at

M. Karner
e-mail: michael.karner@v2c2.at

A. Dorri · S.S. Kanhere
School of Computer Science and Engineering (CSE), University of New South Wales (UNSW), Sydney, Australia
e-mail: alidorri.ce@gmail.com

S.S. Kanhere
e-mail: salil.kanhere@unsw.edu.au

K. Römer
Institute for Technical Informatics, Graz University of Technology, Graz, Austria
e-mail: roemer@tugraz.at

R. Jurdak
Commonwealth Scientific and Industrial Research Organisation (CSIRO), DATA61, Brisbane, Australia
e-mail: Raja.Jurdak@data61.csiro.au

C. Zachäus et al. (eds.), *Advanced Microsystems for Automotive Applications 2017*, Lecture Notes in Mobility, DOI 10.1007/978-3-319-66972-4_12

1 Introduction

Future vehicles will utilize wireless communication networks to interact with other vehicles and road users in close proximity, roadside units like traffic lights and overhead displays at motorways, as well as the Internet. Thereby future connected vehicles will become part of the Internet of Things (IoT) and offer a plethora of beneficial services and applications to the users (e.g., the vehicle owner and driver) as well as the vehicle manufacturers (i.e., the OEM) and their suppliers. However, this high degree of connectivity also raises a wide range of new security threats as well as privacy concerns and will require a comprehensive security architecture. The importance of the latter was recently shown by different hackers attacking modern vehicles via their wireless interfaces (Valasek and Miller 2015; Foster et al. 2015).

Wireless over-the-air (OTA) software (SW) updates will be one of the key features of future connected vehicles and will allow adapting or upgrading the functionality of the vehicle or fixing bugs in the embedded SW installed on its electronic control units (ECU) remotely (Hossain and Mahmud 2007; Khurram et al. 2016). Such updates can be very beneficial for both the OEM (i.e., car manufacturer) as well as the end user (i.e., the vehicle owner) as they obviate the need for taking the vehicle to a service center to receive the latest SW version. However, OTA updates are very critical with repect to security as they require full access to the in-vehicle communication system to allow the installation of new SW images on all ECUs of a vehicle.

Because of their high potential and impact, automotive OTA updates have increasingly attracted the attention of the research community. Researchers have proposed various security architectures and concepts allowing trustworthy OTA updates for vehicles (Hossain and Mahmud 2007; Idrees et al. 2011; Nilsson and Larson 2008). In particular, existing work mainly focus on protecting a vehicle from unauthorized access and the injection of malicious SW. Other authors propose methods allowing secure and efficient SW updates performed locally in a service center or during vehicle assembly Steger et al. (2016), (2016). However, none of the aforementioned solutions address secure OTA distribution of SW from the OEM to all concerned vehicles while ensuring the privacy of the end user. Such a SW distribution process requires a highly scalable security architecture protecting the confidentiality as well as the integrity of the transferred data and furthermore retaining the privacy of the involved users.

In this paper, we propose an automotive security architecture utilizing Blockchain (BC) to tackle the implicated security and privacy challenges of future connected vehicles. Our BC-based security architecture can be utilized to perform OTA updates for smart vehicles remotely as well as to securely distribute the latest SW to service centers or vehicle assembly lines where the latest SW image is installed on the ECU of a vehicle locally. The proposed architecture ensures a secure as well as tamper-proof data exchange and protect the privacy of the end user. Thus, our architecture is not only applicable for protecting wireless automotive SW updates, but can also be utilized in a more general manner to secure a wide range of (future) automotive services.

The proposed architecture is evaluated using a proof-of-concept implementation of a wireless SW update system providing a secure as well as efficient communication between all involved parties: the SW provider (e.g., an automotive supplier) creating the latest SW, the OEM verifying, adapting (e.g., to fit specific vehicle variant) and finally distributing the SW, the cloud storage where the SW is stored, local SW update providers such as a service center, and the connected vehicle itself. We use this implementation to (i) show the applicability of our BC-based architecture for wireless automotive SW updates, (ii) analyze the packet overhead of the architecture due to the use of BC, (iii) highlight its advantages compared to centralized (e.g., certificate-based architecture), and (iv) evaluate the added latency compared to locally performed wireless SW updates.

2 Background

In this chapter, we present an overview of wireless SW updates and thereby describe the technical process as well as the scenarios where these updates can be most beneficial. Furthermore, we give some technical insights on BC, its initial usage as essential part of the crypto-currency Bitcoin, and explain required adaptions to use it in typical IoT as well as automotive applications.

2.1 Wireless Automotive Software Updates

A smart vehicle consists of dozens of ECUs performing different tasks such as controlling the window lifters, the engine, the windscreen wipers, etc. These ECUs are interconnected to each other via the in-vehicle communication system realized using different wired buses (e.g., CAN or LIN). A central vehicle gateway (CGW) is used to interconnect these bus systems as shown in Fig. 1. To wirelessly communicate with the vehicle and all its integrated ECUs, the vehicle has to be equipped with a Wireless Vehicle Interface (WVI) allowing full access to the in-vehicle communication system. Hence, a WVI is also required when a new SW version should be installed on one of the ECUs of a vehicle.

To perform a wireless SW update, the so-called Diagnostic Tester (DT), which possess all required keys to authorize the SW update and the new SW version, connects to the vehicle using its WVI. In the next step, the DT can use automotive diagnostic protocols such as Unified Diagnostic Services (UDS) to (i) initialize and authorize the SW update, (ii) transfer the binary to the ECU, and (iii) install it on the control unit. This procedure can be used locally in a service center, where the DT and the vehicle with a connected WVI are interconnected using a wireless local area

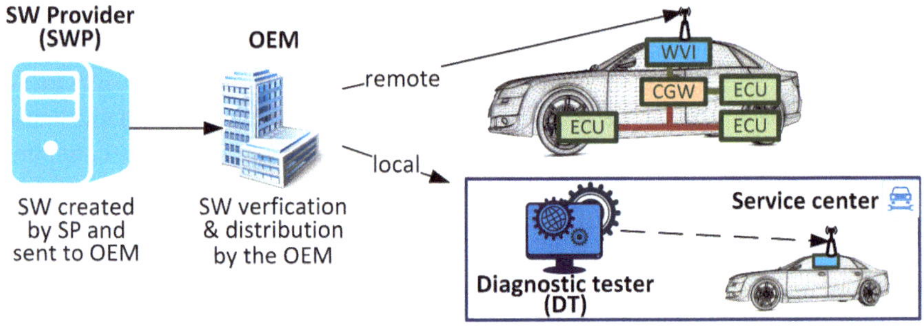

Fig. 1 A new SW version is created by the SW provider (SWP), verified and distributed by the OEM and finally installed on the concerned ECU of a vehicle

network such as Wi-Fi, but also remotely, where a diagnostic service of the OEM communicates with the WVI via the Internet. Both scenarios are sketched in Fig. 1. Local SW updates will mainly take place in service centers during vehicle maintenance, in the vehicle assembly line, and during vehicle development, where engineers will test and compare new SW versions.

A SW update for a vehicle already out in the field (i.e., already sold) is often required due to a (critical) bug in the automotive SW installed on one of its ECUs. The SW of an automotive ECU is often implemented by a supplier company and not by the OEM itself. Thus, in case of a bug fix, the latest (i.e., fixed) SW version is created by the supplier, hereinafter referred to as SW Provider (SWP), then sent to the OEM, and finally distributed to all concerned vehicles. The SW update procedure can either be directly handled between the OEM and the vehicle, or the latest SW is first sent to a service center, where the SW is then installed on the ECU locally. The described example is also sketched in Fig. 1.

2.2 *Blockchains*

BC technology was first introduced in 2008 as essential part of Bitcoin (Nakamoto 2008), the first cryptocurrency network. Since then, BC have been broadly used in nonmonetary applications (e.g., healthcare data exchange (Yue et al. 2016)) due to its security, privacy, and decentralization features. The secure nature of BC originates from the consensus algorithm employed for appending new blocks into the BC. The privacy of the involved users is ensured by utilizing changeable public keys (PK) representing the user. The BC is managed in a distributed fashion by all participating nodes which form an overlay network, thus not requiring any central management.

In a classical BC system the integration of a new block into the BC is done by the consensus algorithm employed to solve computational- and/or memory-expensive, cryptographic puzzles. Such classical systems, however, suffer from some significant limitations mainly caused by the extensive consensus algorithm, namely: (i) high resource consumption, and (ii) high latency.

These limitations are especially critical for a broad range of embedded applications, IoT services, and also resource-constraint ECUs. To tackle the aforementioned limitations, we developed LSB, a Lightweight Scalable BC (Dorri et al. 2017). In LSB, the resource-expensive consensus algorithm is replaced by a timing-based algorithm. Furthermore, LSB divides the network into clusters, which distributedly manage the public BC. Each cluster consists of numerous cluster members (CM) and is managed by one cluster head (CH). Each CH maintains a local copy of the BC and is interconnected to other CHs by the overlay network.

The BC consists of chained blocks containing different and application-specific transactions. To chain the blocks, the newest block contains the hash of the last chained block. A new block is created once the running pool, a local data structure of the CH containing new transactions (i.e., not included in a block yet), has reached a predefined size. In the next step, the new block is broadcasted to the other CHs using the overlay network and finally added to the BC of the other CHs. More detailed descriptions on this process can be found in Dorri et al. (2017).

3 Architecture Enabling Wireless Software Updates

Wireless automotive SW updates must be performed in a secure and dependable way as failed, malfunctioning, or malicious updates will significantly influence the operation of the vehicle and therefore the safety of the passengers. Besides the safety aspect with respect to the involved users (i.e., vehicle driver and owner), privacy considerations are also relevant in certain SW update scenarios (especially remote updates). A suitable automotive security architecture must ensure that a vehicle can receive the latest SW for its ECUs without exposing unrelated personal information about the vehicle and its users. Furthermore, such an architecture must protect the exchanged data at any time to (i) keep required (authorization) keys, mainly required to unlock the ECU, secret, (ii) maintain the confidentiality of the SW image, and (iii) ensure the integrity of the transferred data to avoid manipulation. These requirements are valid for the entire chain shown in Fig. 1: first the image is sent from the SWP to the OEM, second the SW image is forwarded to concerned vehicles and local SW update providers, and third the image is installed on the ECU (e.g., using a local wireless network in a service center).

While the local update scenario is already covered by security concepts such as Steger et al. (2016), the security of the SW distribution from a SWP to an OEM and then further to the vehicles is still an open issue. Some OEMs like Tesla (Gabe 2016) currently perform OTA updates using VPN tunnels between the OEM server infrastructure and the vehicle itself. Although this approach is suitable to protect the transferred data, it also requires a dedicated point-to-point link between the OEM and the vehicle which can potentially be critical with respect to the privacy of the end user. Other automotive security architectures use certificates to establish trust within the network (Woo et al. 2016; Aslam and Zou 2009) or (Mallissery et al. 2014). However, we believe that these centralized approaches are not suitable for highly distributed scenarios encompassing thousands of vehicles around the globe and therefore propose to use a BC-based automotive security architecture instead.

3.1 *Blockchain-Based Architecture Securing Wireless Software Updates*

Our BC-based architecture described in this section is able to fulfill the aforementioned security, privacy, and scalability requirements. In the following, we will focus on the secure SW image distribution from the SWP and the OEM to the target vehicles as shown in Fig. 1. Our architecture presented in Fig. 2, protects the transfer of SW images and also ensures the privacy of the involved users.

Our architecture is based on the design presented in Dorri et al. (2017). However, several adaptations were required to ensure that our architecture can meet the needs of the automotive domain. Additionally, different stakeholders and roles are involved in automotive scenarios compared to typical smart home applications.

As shown in Fig. 2, these roles are SW providers, OEMs, cloud storages (CS), and vehicular interfaces (VI) representing either smart connected vehicles or local SW update providers such as service centers or vehicle assembly lines.

The cloud storage is essential for our architecture as it serves as a repository for storing new SW images sent by the SWP or the OEM, and furthermore handles the data download to the VIs. In our architecture, the cloud storage provides a sophisticated authentication mechanism to ensure that only authorized entities can write, adapt, and download a specific SW image.

All aforementioned entities are interconnected using the overlay network. This clustered network allows unicast as well as broadcast data streams and is able to

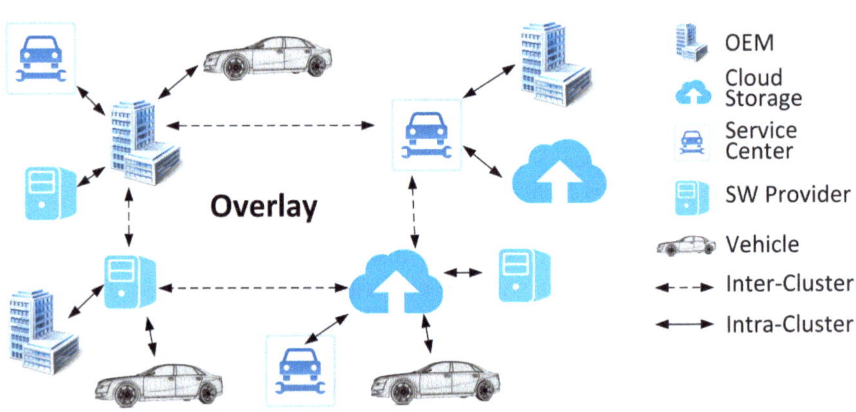

Fig. 2 The proposed BC-based architecture to securely interconnect all involved parties

provide suitable message flows for all different data exchanges involved in the SW distribution process. As described in Sect. 2.2, the overlay network basically interconnects different (e.g., geographical) clusters encompassing one CH and numerous CMs. Thereby, a CH acts as gateway for messages sent from a CM of a specific cluster (e.g., CM1 in cluster A) to CM or CH of another cluster (e.g., CM2 in cluster B). In our architecture, a CM and a CH can occupy specific roles such as acting as OEM, CS, or VI. However, as a CH besides its dedicated role also has to maintain its local BC, it is very unlikely that a vehicle would act as a CH due to its resource constraints and the fact that a vehicle is mobile and therefore not able to be connected to the overlay at all times. Our architecture uses two types of messages: (i) *transactions* and (ii) *blocks* for implementing OTA updates.

Transactions are used to initialize the BC system (*genesis transaction)* and to handle the SW distribution process (*update transaction*). The latter is required to inform the OEM (i.e., when sent by the SWP) as well as the concerned VIs (i.e., when sent by the OEM) about newly available SW. Update transactions, thus, play a vital role in our architecture as they contain required information for both the OEMs and the VIs: it contains the identities of the SWP as well as of the OEM (i.e., their PKs and signatures), the transaction ID (i.e., hash representation of the transaction), the ID of the previously created transaction (the genesis transaction for the very first update transaction), and the metadata field. The metadata field is used by the SWP and the OEM and includes information about the SW update itself as well as the location where the new image is stored.

A BC-block consists of several transactions as well as a link to the previous block that chain these blocks together. It is created every time the running pool of a CH reaches a predefined size. Each block has a unique ID and is signed with the private key of the creating CH. After creation, the block is broadcasted to all other CHs of the overlay for verification and then chained to the local BC.

3.2 Employing Our Architecture to Distribute New SW

The proposed automotive security architecture can be employed to securely distribute a new SW image to the target vehicles. We will now sketch the corresponding process, explain all involved steps and utilized message types, and show that our architecture is able to protect the entire process and all involved users.

In the vehicle assembly, the OEM will store its PK on each assembled vehicle and the vehicle will generate a secret key pair (e.g., an RSA key pair consisting of a private and a public key). Both the PK of the OEM and the key pair will be securely stored on the WVI in a tamper-proof storage. While a vehicle is assembled it can also create a genesis transaction, an initial transaction required to participate in the BC. This process can be highly OEM-dependent and will therefore not be described in more detail. However, we suggest that this transaction include information about the vehicle type (e.g., vehicle variant) and that the transaction is signed by the OEM. Or, as an alternative, a dedicated token including the aforementioned data and the signature of the OEM is created at this point in time. This transaction/token is required later to request new SW stored on a cloud storage.

A SW distribution process is triggered by a SWP when a new image is created potentially due to a necessary bug fix. Once the new SW is developed, the SWP will create a store request including the signature of the SWP and send it to the cloud storage. The latter will verify the request, locally initialize the process, and send a *store response* including the signature of the storage and a file descriptor required as reference for the data upload process back to the SWP. Please note that the SW can also be created by the OEM itself. In this case the OEM would upload the SW to the CS. The rest of the process is similar to that described above, except that the OEM will take over the tasks performed by the SWP. Once the data is stored on the cloud storage, the SWP creates an update transaction including information about the location of the image on the cloud storage, adds the PK of the concerned OEM, signs the transaction with its private key and finally broadcasts the transaction to the overlay network. Please note that the transaction is not valid yet as the second signature is missing and therefore it is not added in the running pool of the CHs.

In the next step the OEM receives the update transaction, verifies it, validates, and if required adapts the SW image stored on the cloud storage, and finally also signs the update transaction. The transaction which is now valid is again broadcasted to the overlay and locally stored by the CHs. The CHs will also send the transaction to all CMs in its cluster to inform vehicles about the new SW.

Finally, the valid transaction is received by the target VIs (i.e., vehicles or local SW update providers). After validating the transaction and parsing the metadata, the VIs will send a signed *download request* including the token signed by the OEM (e.g., stored on the WVI when a vehicle was assembled) to the cloud storage to receive the new SW version. The CS will validate the request, use the token to verify that the vehicle is applicable for the new SW, and is finally utilizing a unicast data stream to send the image to the VI, where the SW is installed.

4 Proof of Concept

In this section, we describe the implementation of the BC-based security architecture presented in the previous section. The implemented framework consists of two main building blocks (Fig. 3). First, the overlay network used by the SWP to send the latest SW to the OEM and employed by the OEM to distribute the verified SW version to the concerned vehicles as well as local distributors (e.g., service centers). Second, the local update process required to install the latest SW on the ECU. This step is executed by the WVI, which receives the latest SW either from the OEM via a remote connection or locally from a DT.

The implementation of the overlay encompasses the development of the CH and the CM as well as a local test suit to set up overlay test topologies (mainly on local host) consisting of several instances of the CHs and CMs. All the above is implemented in Java.

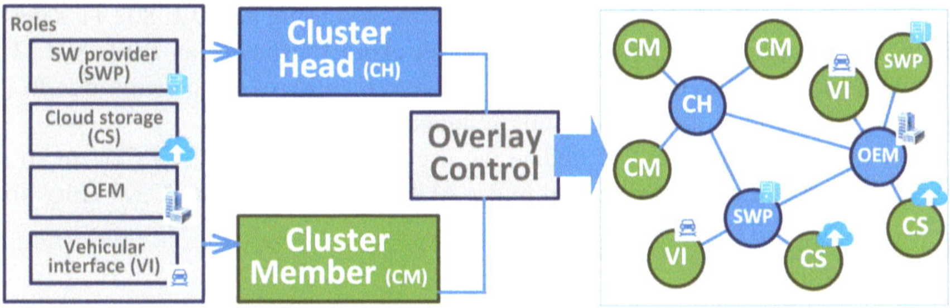

Fig. 3 Proof-of-concept implementation encompassing prototypes of the CHs and the CMs. The overlay control allows to design different test scenarios by utilizing serval CH and CM instances, assigning specific roles to these instances and defining the topology of the overlay

Our implementation allows us to evaluate different scenarios for different overlay topologies and numbers of VIs. We have also implemented a baseline system which is similar to the state of the art, wherein a dedicated Certificate Authority (CA) is employed to verify certificates used by the SWP as well as the OEM.

The local SW update building block is based on the wireless automotive SW update framework we presented in Steger et al. (2016). This framework allows to perform local SW updates in a secure and dependable way by employing IEEE 802.11 s to interconnect the DT and the vehicle with an integrated or connected WVI (via OBD).

Our WVI prototype, as shown in Fig. 4a, consists of a BeagleBone Black board (BBB), an additional communication cape (i.e., a printed circuit board allowing the BBB to connect to a vehicle via CAN/OBD), and the corresponding SW implementation (Java and C). The DT prototype is also realized in Java and can therefore be used on a normal PC but also on a BBB. Both prototypes provide different SW update mechanisms and allow different diagnostic functions.

(a) **(b)**

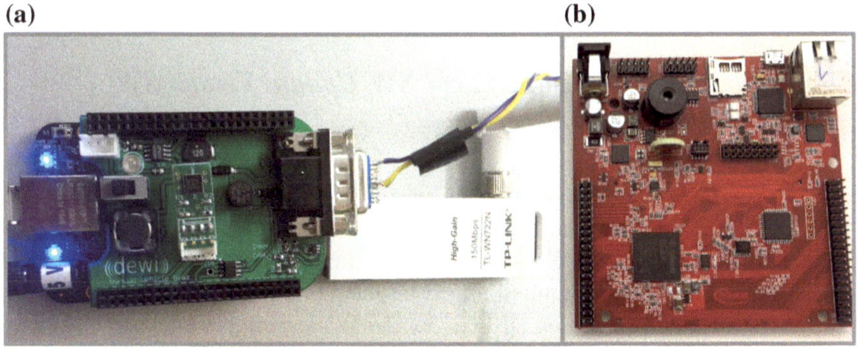

Fig. 4 **a** The WVI prototype based on a BeagleBone Black and our developed communication cape; **b** target ECU: Infineon AURIX ECU in the AURIX application kit TC277 TFT

The vehicles' WVI and the DT are interconnected using an IEEE 802.11 s mesh network. We chose this protocol as the mesh characteristics of an IEEE 802.11 s network increases the flexibility as well as the reliability of the network due to its multi-hop capability and the resulting redundancy.

As target ECU for the SW update we use an Infineon AURIX ECU, an automotive multi-core ECU, assembled in the AURIX application kit TC277 (Fig. 4b).

5 Evaluation

We used our proof-of-concept implementation to show the applicability of our architecture to fulfill the needs of an automotive OTA SW update system. Thereby, we evaluated the (packet) overhead when using BC and compared the duration of the BC-based SW distribution with the time required to install a SW update locally on an ECU. Furthermore, we compared our architecture with the baseline certificate-based system outlined in Sect. 5. We used this evaluation to compare the total number of packets exchanged as well as the latency incurred in the SW distribution process.

5.1 Overhead Due to the Use of Blockchains

In the first evaluation step we analyzed the overhead added by the BC. Therefore, we collected the number of exchanged packets and grouped them into data-related, BC-related, and packets required to initialize the system. The overhead is affected by the number of VIs, the size of the binary, and the number of performed updates. Our evaluation for a 32 KB binary and 100 SW updates per VI reveals an added overhead of 3.4% for 20 VIs connected to the overlay and up to 7.3% if only 1 VI is updated. Neglecting the initialization packets, the overhead is only 3.3% for twenty VIs.

5.2 Latency Comparison: Local SW Update Versus SW Distribution Using BC

In this experiment, we compared the latency added by the BC-based SW distribution and the latency of the SW update process itself. For this, we measured the latency of a local wireless SW update using the framework presented in Steger et al. (2016) as well as the time required for the last step of a remote SW update, where a new SW update is installed on an ECU by the WVI using the vehicle bus system.

The results are presented in Table 1 and show that the installation of a new image on the ECU using the wired in-vehicle bus system takes more than five times

Table 1 Comparison of the latency of the SW distribution, a local wireless SW update, and the installation of a new SW image on an ECU performed by the WVI

SW distribution	Wireless local update	WVI installation
2682.3 ± 8.3 ms	16271.0 ± 323.4 ms	13831.7 ± 228.3 ms

longer than the SW distribution from an emulated SWP to the VI using our BC-based system and that the SW distribution process using our experimental setup is about six times faster than the local SW update process. Note that the SW distribution was performed using a LAN network architecture. Therefore, the SW distribution latency does not include any additional latency caused by typical Internet links. The latency of local SW update and the SW installation can also vary depending on the used ECU, the employed security mechanisms, etc.

5.3 Comparison of BC- and Certificate-Based Approaches

We evaluated the number of exchanged packets as well as the latency of our BC-based system compared to a certificate-based approach. Therefore, we used an overlay network consisting of up to ten HW nodes (several BBBs, Raspberry Pi3's, and a Laptop) interconnected by the overlay network and performed measurements using different network topologies and different numbers of VIs.

The evaluation results presented in Figs. 5 and 6 show that both approaches have quite similar properties with respect to the added latency as well as the total number of exchanged packets and that our BC-based approach is slightly better than certificated-based approach in both aspects.

The performed experiments and measurements showed that (i) BC approximately add 3% packet and 14% latency overhead compared to a pure OTA SW update, (ii) BC has lower latency and uses fewer packets than a certificate-based system, and (iii) show the applicability of our architecture for automotive applications.

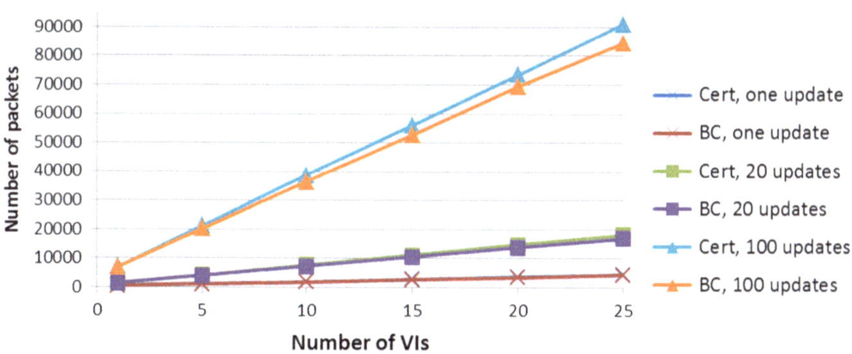

Fig. 5 Exchanged packet count comparison with respect to the number of VIs and number of updates

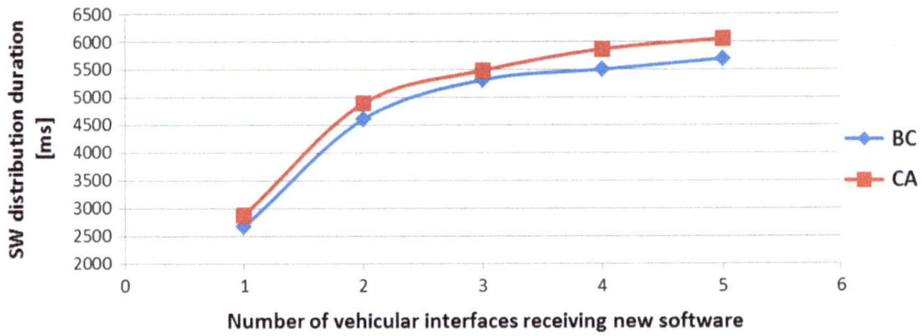

Fig. 6 Comparison of the SW distribution duration for different numbers of involved VIs

6 Conclusion

In this paper, we proposed a security architecture based on BC for smart-connected vehicles able to support a broad range of (future) automotive applications and services. Our architecture provides a secure and trustworthy interconnection between all involved parties while ensuring the privacy of the involved users. We evaluated our architecture using a proof-of-concept implementation of a wireless SW update system and use the latter to show the applicability of our architecture as well as its benefits compared to a certificate-based system. We plan to further refine our architecture, to improve our implementation, and to perform more detailed evaluations employing more nodes in the next months.

References

Aslam B, Zou C (2009) Distributed certificate and application architecture for VANETs. In: IEEE military communications conference, pp 1–7

Dorri A, Kanhere S, Jurdak R (2017) Towards an optimized blockchain for IoT. In: Proceedings of the second international conference on internet-of-things design and implementation (IoTDI '17). ACM, pp 173–178

Foster D, Prudhomme A et al (2015) Fast and vulnerable: a story of telematic failures. In: USENIX workshop on offensive technologies

Gabe N (2016) Over-the-air updates on varied paths, automotive news

Hossain I, Mahmud S (2007) Analysis of a secure software upload technique in advanced vehicles using wireless links. In: Intelligent Transportation Systems Conference, pp 1010–1015

Idrees M, Schweppe H et al (2011) Secure automotive on-board protocols: a case of over-the-air firmware updates. Lecture Notes in Computer Science. LNCS, vol 6596, pp 224–238

Khurram M, Kumar H et al (2016) Enhancing connected car adoption: security and over the air update framework. In: IEEE world forum on internet of things (WF-IoT), vol 3, pp 194–198

Mallissery S, Pai M et al (2014) Improving the PKI to build trust architecture for VANET by using short-time certificate mngt. and Merkle Signature Scheme. In: Asia-Pacific conference on computer aided system engineering, pp 146–151

Nakamoto S (2008) Bitcoin: a peer-to-peer electronic cash system. http://www.bitcoin.org/bitcoin.pdf
Nilsson D, Larson U (2008) Secure firmware updates over the air in intelligent vehicles. In: IEEE conference on communications, pp 380–384
Steger M, Karner M et al (2016) Generic framework enabling secure and efficient automotive wireless SW updates. In: IEEE international conference on emerging technologies and factory automation (ETFA), vol 21, pp 1–8
Steger M, Karner M et al (2016) SecUp: secure and efficient wireless software updates for vehicles. In: IEEE conference on digital system design (DSD), pp 628–636
Valasek C, Miller C (2015) Remote exploitation of an unaltered passenger vehicle, White Paper, p 93
Woo S, Jo H et al (2016) A practical security architecture for in-vehicle CAN-FD. IEEE Trans Intell Transp Syst 17:2248–2261
Yue X, Wang H et al (2016) Healthcare data gateways: found healthcare intelligence on blockchain with novel privacy risk control. J Med Syst 40:1–8

DEIS: Dependability Engineering Innovation for Industrial CPS

Eric Armengaud, Georg Macher, Alexander Massoner, Sebastian Frager, Rasmus Adler, Daniel Schneider, Simone Longo, Massimiliano Melis, Riccardo Groppo, Federica Villa, Padraig O'Leary, Kevin Bambury, Anita Finnegan, Marc Zeller, Kai Höfig, Yiannis Papadopoulos, Richard Hawkins and Tim Kelly

Abstract The open and cooperative nature of Cyber-Physical Systems (CPS) poses new challenges in assuring dependability. The DEIS project (Dependability Engineering Innovation for automotive CPS. This project has received funding from the European Union's Horizon 2020 research and innovation programme under grant agreement No 732242, see http://www.deis-project.eu) addresses these challenges by developing technologies that form a science of dependable system integration. In the core of these technologies lies the concept of a Digital Dependability Identity (DDI) of a component or system. DDIs are modular, composable, and executable in the field facilitating (a) efficient synthesis of component and system dependability information over the supply chain and (b) effective evaluation of this information in-the-field for safe and secure composition of highly distributed and autonomous CPS. The paper outlines the DDI concept and opportunities for application in four industrial use cases.

E. Armengaud (✉) · G. Macher · A. Massoner · S. Frager
AVL List GmbH, Hans List Platz 1, 8020 Graz, Austria
e-mail: eric.armengaud@avl.com

G. Macher
e-mail: Georg.macher@avl.com

A. Massoner
e-mail: alexander.massoner@avl.com

S. Frager
e-mail: sebastian.frager@avl.com

R. Adler · D. Schneider
Fraunhofer Institute for Experimental Software Engineering (IESE), Fraunhofer Platz 1, 67663 Kaiserslautern, Germany
e-mail: rasmus.adler@iese.fraunhofer.de; rasmus.adler@iese.fhg.de

D. Schneider
e-mail: daniel.schneider@iese.fraunhofer.de; daniel.schneider@iese.fhg.de

S. Longo · M. Melis
General Motors—Global Propulsion System, Corso Castelfidardo, 36, 10138 Turin, Italy
e-mail: simone.longo@gm.com

M. Melis
e-mail: massimiliano.melis@gm.com

C. Zachäus et al. (eds.), *Advanced Microsystems for Automotive Applications 2017*, Lecture Notes in Mobility, DOI 10.1007/978-3-319-66972-4_13

Keywords Safety · Security · Digital Dependability Identity (DDI) · Automotive · Railways · Healthcare

1 Introduction

Cyber-Physical Systems (CPS) harbor the potential for vast economic and societal impact in domains such as mobility, home automation, and delivery of health. At the same time, if such systems fail they may harm people and lead to the temporary collapse of important infrastructures with catastrophic results for industry and society. CPS is the key to unlocking their full potential and enabling industries to develop confidently business models that will nurture their societal uptake. Using currently available approaches, however, it is generally infeasible to assure the dependability of Cyber-Physical Systems. CPS are typically loosely connected and

R. Groppo
Ideas & Motion SRL, Via Moglia 19, 12062 Cherasco (CN), Italy
e-mail: riccardo.groppo@ideasandmotion.com

F. Villa
Politecnico di Milano, Piazza Leonardo da Vinci, 32, 20133 Milan, Italy
e-mail: federica.villa@polimi.it

P. O'Leary · K. Bambury
Portable Medical Technology Ltd, Google Campus, 4-5 Bonhill St, London EC2A 4BX, UK
e-mail: padraig@portablemedicaltechnology.com

K. Bambury
e-mail: kevin@portablemedicaltechnology.com

A. Finnegan
RSRC at Dundalk Institute of Technology, Marshes Upper, Dundalk, Co. Louth, Irland
e-mail: anita.finnegan@dkit.ie

M. Zeller · K. Höfig
Siemens AG, Otto-Hahn-Ring 6, 81739 Munich, Germany
e-mail: marc.zeller@siemens.com

K. Höfig
e-mail: kai.hoefig@siemens.com

Y. Papadopoulos
University of Hull, Cottingham Rd, Hull, Hull HU6 7RX, UK
e-mail: Y.I.Papadopoulos@hull.ac.uk

R. Hawkins · T. Kelly
University of York, Heslington, York YO10 5DD, UK
e-mail: richard.hawkins@york.ac.uk

T. Kelly
e-mail: tim.kelly@york.ac.uk

come together as temporary configurations of smaller systems which dissolve and give place to other configurations. The key problem in assessing the dependability of CPS is that the configurations a CPS may assume over its lifetime are unknown and potentially infinite. State-of-the-art dependability analysis techniques are currently applied during the design phase and require a priori knowledge of the configurations that provide the basis for the analysis of systems. Such techniques are not directly applicable, can limit runtime flexibility, and cannot scale up to CPS.

The DEIS project addresses these important and unsolved challenges by developing technologies that form a science of dependable system integration. In the core of these technologies lies the concept of a Digital Dependability Identity (DDI Schneider et al. 2015) of a component or system. The DDI targets (1) improving the efficiency of generating consistent dependability argumentation over the supply chain during design time, and (2) laying the foundation for runtime certification of ad hoc networks of embedded systems. Main contributions of this paper are the introduction of the DDI concept and opportunity analysis for the use of DDI in four relevant industrial use cases from three different domains. The paper is organized as follows: Sect. 2 introduces the DDI concept, while in Sect. 3 the four industrial use cases are presented. In Sect. 4, the opportunities for using the DDI are summarized, and finally Sect. 5 concludes this paper.

2 The Digital Dependability Identity (DDI) Concept

In general, a Digital Identity is defined as "the data that uniquely describes a person or a thing and contains information about the subject's relationships" (Windley 2005). Applying this idea, a DDI contains all the information that uniquely describes the dependability characteristics of a system or component. This includes attributes that describe the system's or component's dependability behavior, such as fault propagations, as well as requirements on how the component interacts with other entities in a dependable way and the level of trust and assurance, respectively. In general, A DDI is a living model-based modular dependability assurance case. It contains an expression of dependability requirements for the respective component or system, arguments of how these requirements are met, and evidence in the form of safety analysis artifacts that substantiate arguments. A DDI is produced during design, issued when the component is released, and is then continually maintained over the complete lifetime of a component or system. DDIs are used for the integration of components to systems during development as well as for the dynamic integration of systems to "systems of systems" in the field.

A prerequisite is the availability of a common and machine-readable communication language which shall be independent of specific development approaches and tools, and finally enables collaboration between actors in the value chain. Although progress has been achieved with dependability metamodels, e.g., within architecture description languages like EAST-ADL and AADL, there is still a lack of a common model for the communication of dependability information. The recently released (Structured assurance case metamodel SACM 2016) defines a metamodel for representing structured assurance cases, i.e., a set of auditable claims, arguments, and evidence created to support the claim that a defined system/service will satisfy the particular requirements. The SACM will be the metamodel for the externally visible DDI interface, whereas the internal "logic" will be described by other, already existing approaches.

Managing the variability of a system is a key challenge in the industrial context, e.g., during different development iterations or for the management of product variants. This requires that DDIs support change impact analyses that enable a prediction about whether a component will fit into multiple product variants, to help reduce re-assurance effort. As these models would be modularized in the level of components, DDI would allow to conduct such analyses across different DDI employing different techniques (such as C2FTs or Hip-HOPS Azevedo et al. 2014; Dheedan and Papadopoulos 2010).

In connected CPS, dependability cannot be fully assured prior to deployment, because systems will dynamically interconnect (with third-party systems) and form systems of systems with largely unpredictable consequences for dependability. In order to assure the dependability of such in-field integrations, we propose automated DDI-based dependability checks at integration time. DDIs, therefore, must become executable specifications accompanying systems through their complete life cycle, not simply digital artifacts that cease in their utility after deployment of the system.

A central question is, however, to which extent "safety intelligence" can be shifted from development time into runtime (see Fig. 1). A possible first step is to use runtime certificates such as ConSerts (Schneider and Trapp 2013). In this case, the assurance case is still completely designed and managed at development time, only series of unknown context-dependencies are formalized and shifted into runtime, "known unknowns". The next step would be to deal with "unknown unknowns" at the design level, with the requirements and thus the safety goals still

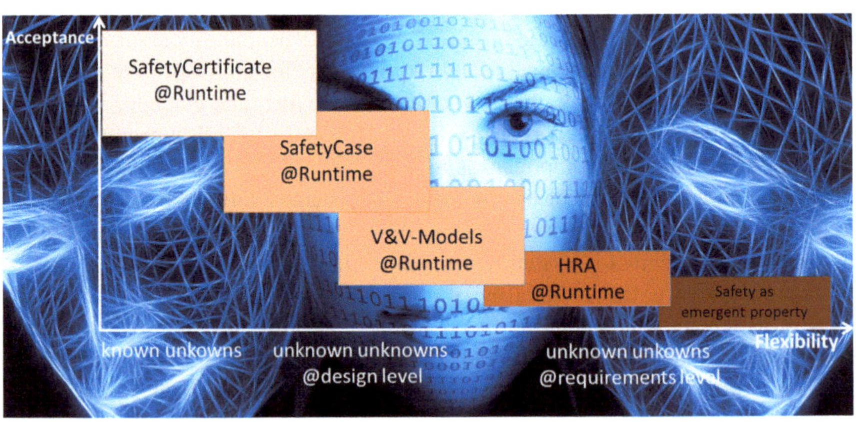

Fig. 1 The different levels of safety variability

being determined by human engineers. While SafetyCases@Runtime would be capable to run predefined validation and verification activities at runtime to obtain required evidences, V&V-Models@Runtime additionally support the modification of V&V-models, e.g., the modification of test cases or pass/fail criteria. As soon as the adaptation to "unknown unknowns" also requires an adaptation of requirements, it is additionally necessary to adapt the hazard and risk analysis and the resulting safety goals at runtime. With that last step, a crosscutting aspect of the whole dependability life cycle would be shifted into runtime, which is today hardly conceivable.

To achieve this vision, in DEIS we develop DDIs as modular, composable, and executable model-based assurance cases with interfaces expressed in SACM and internal logic that may include fault trees, state automata, Bayesian networks, and fuzzy models for representing uncertainty. Modularity means DDIs apply to units at different levels of design, including the system itself, its subsystems, and components. By being composable, the DDI of a unit can be derived in part from its constituent elements. To simplify the DDI construction and synthesis, in DEIS we experiment with techniques for automatic, model-based construction of DDIs via automatic allocation of safety requirements and auto-generated model-based dependability analyses (Azevedo et al. 2014). To enable executability, we explore the use of the models that represent the internal logic of DDIs for dynamic detection and prognosis of risks, considering among other factors limitations in observability and environmental uncertainties (Schneider and Trapp 2013; Dheedan and Papadopoulos 2010). Within the scope of DDI modeling, we include examination of implications of security breaches on safety.

3 The Four Industrial Use Cases in DEIS Project

3.1 Automotive: Development of a Stand-Alone System for Intelligent Physiological Parameter Monitoring

Global healthcare is worth 1.5$ trillion, and a part of that, around 11$ billion, is focusing on physiological portable monitor system (e.g., wearable and quantified self) (HRI analysis and centers for medicare and medicaid services national health expenditures 2012; PwC health research institute, new health entrants 2015). Moreover, a recent New AAA Foundation report reveals drivers spend an average of more than 17,600 min behind the wheel each year. Future frontiers of smart mobility rise the need to measure the physiological parameter of the driver and passengers and to evaluate the health of the occupants. This to enable autonomous driving features according to driver health state, and enable potential B2B and B2C opportunities.

DDIs are a revolutionary concept for the automotive domain and can open up new prospects, providing means for ensuring dependability among CPS and

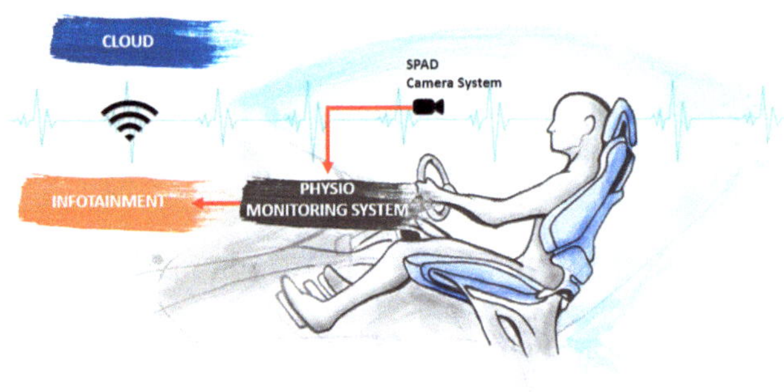

Fig. 2 Intelligent physiological parameter monitoring—overview

improving the safety of driver and passengers, taking actions (supporting with autonomous drive and/or health services) in case of emergency/needs. The aim of this use case is the introduction of a dependable physiological monitoring system applied to a connected vehicle, capable to identify physiological parameters and to evaluate the health of driver and other occupants. The proposed environment (see Fig. 2) contains a comprehensive package of technologies, tools, and services that support the drive session in evaluating driver health state, and enable autonomous driving features according to the needs. The same acquired parameters are also transmitted to a cloud-based system for online support and health analysis which is provided back to the customer as additional service.

While there is an opportunity to improve safety, at the same time the system is potentially prone to a security attack. One of the challenges will be to maximize the *Confidentiality* of health data transmitted, and guarantee *Safety* and *Plausibility* analysis on acquired signals, as well as *Integrity* on communication among different CPSs.

Application of DDIs shall help protection of sensitive information sent through the network, granting the authenticity on actors involved, senders, and receivers. Crypto methodologies should be used to avoid tampering and prevent corruption during transmission. The DDI shall also help to guarantee a certain level of privacy on the sensitive physiological data collected and transmitted. The cloud-based system shall collect all data of each CPS and keep data in sync for each session. Overall, the system shall be capable to access all the services and send/receive DDIs in any operative and network conditions. The physiological monitor system shall be representative of the true driver health condition and noise resistant. This is because, depending on the driver health condition, the system should react promptly activating a predefined action, minimizing delays or applying work-arounds to possible network unavailability. The physical system should remain compliant with the automotive standard (ISO 26262) in the case that autonomous drive action is enabled.

3.2 Automotive: Enhancement of an Advanced Driver Simulator for Evaluation of Automated Driving Functions

This use case introduces a novel driver simulator based on AVL VSM[1] for the comprehensive evaluation of complex autonomous driving functions. The proposed environment contains an extensive package of tools and services that support the OEM in the prediction of vehicle behavior and enables improvement of various vehicle attributes from the initial concept to the testing phase.

An approach for the optimal control of a fully electric vehicle and its powertrain approaching a road segment with Multiple Traffic Lights (TL) has been presented in Ferreira-Parrilla et al. (2014). A system referred to as the Traffic Light Assistant (TLA) was developed to in order to take over longitudinal control of the vehicle, optimizing the velocity trajectory when approaching multiple traffic lights in traffic. The main aims of the system are to reduce energy consumption and CO2 emission, reduce the number of vehicle stops and waiting times, and introduce smoother speed profiles (see Fig. 3). The presented TLA approach has since been further developed to work with other powertrain topologies (Jones et al. 2016a, b) and was demonstrated in real time for conventional vehicles on the powertrain testbed within the FFG Austrian Funded R&D Project TASTE (Traffic Assistant Simulation and Testing Environment).

Previous work has focused on controlling an "ego vehicle" equipped with TLA, assuming complete knowledge about the road conditions and traffic light signal phasing. In reality, not all information may be accessible at all times to every vehicle. For example, if the current state of a traffic light or a pedestrian at the side of a crossing is observed by an onboard camera or local V2I communication, the information would only be available to a vehicle close by. Such a vehicle, if equipped with automated driving functions, would adapt its velocity accordingly to stop in front of the traffic light, or to safely let the pedestrian pass. However, the following vehicle might not directly see the traffic light or the pedestrian and therefore could not anticipate the behavior of the preceding vehicle. In this case, a reaction can only occur based on observation of the behavior of the preceding vehicle. It is possible to improve energy efficiency, if the state of traffic light or the presence of pedestrian are known in advance.

Once more the opportunity for increased efficiency and safety comes with increased security threats. Thus, DDIs will explore how to improve control strategies to increase energy efficiency while ensuring safety at the same time. Equally important would be to analyze and evaluate the impact of possible security attacks, and further define strategies to avoid these attacks or mitigate their effects (Armengaud et al. 2015).

---

[1]https://www.avl.com/-/avl-vsm-vehicle-simulation.

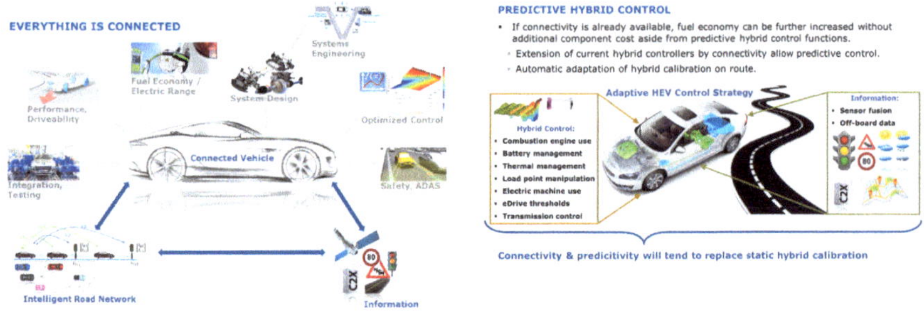

Fig. 3 The connected powertrain and autonomous driving functions allow increased energy efficiency (reduction of fuel consumption, emissions)

3.3 Railway: Enabling Plug-and-Play Scenarios for Heterogeneous Railway Systems

With a total market volume of 47 billion € and nearly 400,000 employees in the European rail industry (by 2014), the rail industry plays a key role in mobility in the EU (UNIFE 2014). The European railway domain has to cope with the challenging situation of heterogeneous systems of systems with different standards and system qualities, e.g. interoperability between train side and track side systems. Moreover, railway systems typically comprise both legacy systems and systems at the cutting edge of technology, which are potentially not designed to be integrated but need to interact in operation.

The European Commission's long-term objective is to achieve a *Single European Railway Area* to deliver the benefits of market opening and interoperability as well as to reduce the life cycle costs (i.e., the costs of developing, building, maintaining, operating, renewing and dismantling) of rolling stock and onboard signaling systems (50% reduction by 2030) (EC 2011). The challenges to reach this goal not only include harmonizing technical interfaces, but also a common signaling system.

The European Train Control System (ETCS) provides standardized train control in Europe and eases traveling with trains crossing the borders of all countries in Europe. Due to historical reasons, different trains and also different track side solutions (such as balises, track circuits, or GSM-R transmission) are often present even within one country. To overcome the challenges, the EC fosters the technical harmonization within the European railway sector ensuring interoperability in the railway domain. Thus, enabling "plug & play" of railway systems scenarios as a long-term objective (Shift2Rail 2015). However, in such "plug & play" scenarios guaranteeing system dependability requirements pose new challenges. Consequently, certification activities (with respect to the CENELEC standards EN 50126 & 50129) require accurate planning and must react quickly to changes within the system development process. Systems of systems in the railway domain are also produced by various stakeholders in the value chain (such as national or even regional public transport authorities, national safety authorities, railway undertaking, OEMs, suppliers, etc.) and, therefore, safety information about components and subsystems (rolling stock, track side, and railway systems) need to be interoperable and exchangeable.

As illustrated in Fig. 4, ETCS consists of an onboard and a track side system. Both subsystems must fulfill the safety requirement as defined in the ERTMS/ETCS specification. Thereby, specific hazards as well as tolerable hazard rates are apportioned to each system. Moreover, track side and onboard systems are often provided by different vendors. ETCS onboard and track side systems must meet the specified safety requirements specified in the specification in order to be safely integrated into any interoperable railway system.

In this use case, we use DDIs to interchange safety-relevant information (including, e.g., safety requirements, models, and assessments) during the development life cycle of the track side and the on-board ETCS units. We, thereby, target to demonstrate the improvement of interoperability in the area of safety engineering across companies, railway operators as well as safety authorities. Hence, we show how time and effort in the certification of systems (or subsystems) can be reduced significantly by interchanging and reusing dependability information across the value chain of the railway domain.

3.4 Health Care: Enhancement of Clinical Decision App for Oncology Professional

The healthcare industry globally is moving toward Electronic Health Records (EHR's) for the management and storage of patient data.[2] ONCOassist[3] is a clinical decision support app for oncology professionals, see Fig. 5. It contains all the key oncology decision support tools oncology professionals need and makes them available in an easy to access and interactive format at point of care. It is CE approved, meaning it is classified as a medical device and was developed in a regulated environment using ISO13485 and IEC62304.

The demand for clinical decision support apps is growing rapidly because of (a) genomic sequencing, (b) aging populations, and (c) new targeted therapies. These changes in the market mean the decision-making process for oncology professionals is becoming more difficult and complex and they need systems to support them with this. ONCOassist is addressing this market need by putting key decision support information and tools in the palm of the hand of the oncology professionals. The ability to transfer new dependability-relevant data from service

[2]http://www.healthcareitnews.com/news/precision-medicine-growth-hinges-electronic-health-records.

[3]http://oncoassist.com/.

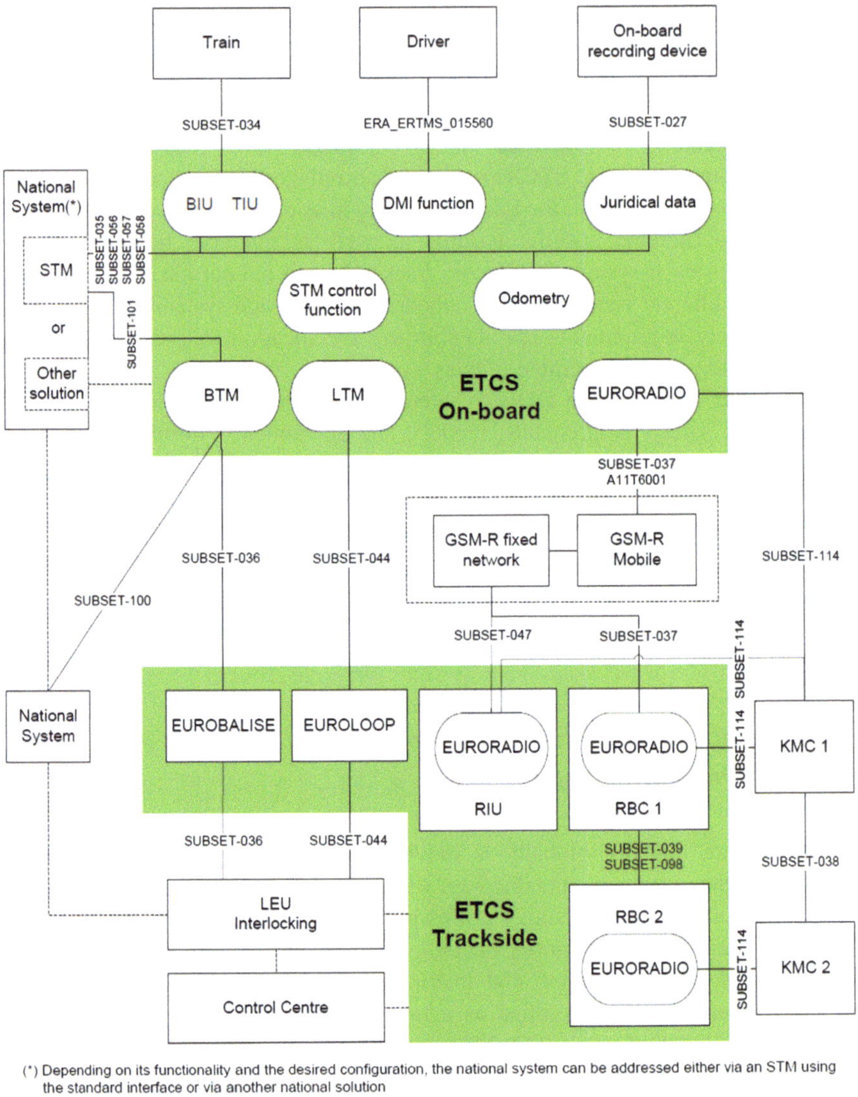

Fig. 4 ERTMS/ETCS reference architecture

provider to a data repository is increasingly important. This new dependability-relevant information could be integrated into an existing data record within the data repository in a way that preserves the intended meaning of the information being transferred. The service provides dependability information in form of DDIs and exchanges them with the data repository. When the systems meet at runtime, they use the DDI's to check if they can provide together new services in a dependable way. If this check is positive, then they start collaborating and

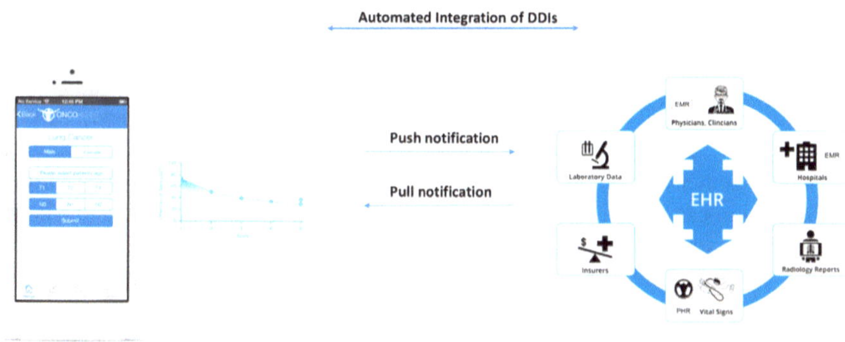

Fig. 5 ONCOAssist—clinical decision support app for oncology professionals

providing the service, exchanging data and information. This, in turn, means that we can provide new services with confidence on the integrity of information handled. DDIs in this context can result in ad hoc (in the wild) integration of ONCOassist with EHRs and hospital information systems, in turn leading to context-aware decision support for clinicians.

4 Opportunities for DDI Applications

As an outcome of the analysis of the use cases presented in Sect. 3, four scenarios for the usage of DDI in industrial application were identified

Sc1: Creation of integrated system assurance case for distributed development of dependable systems
In a supply chain, the OEM provides system dependability requirements which must be taken into account by suppliers. The component supplier engineers create an assurance argument for their component including the definition of assurance claims for the component, evidence relating to those claims and argument explaining how the evidence supports the claims. Information and evidence from the assurance case is translated into an SACM model for exchange as a DDI. This activity may be automated for component and system models used in the assurance case. Once the translation to DDIs is complete the higher level assurance case for the system is created by the OEM via analyzing and integrating information exchanged in the DDIs. This activity may be at least partially automated in order to support dynamic assurance case creation at runtime. Expected result is a strong efficiency increase by compiling and updating complex system assurance cases thanks to the availability of the SACM standard, the DDI framework and the related tools that automate part of the task.

Sc2: Runtime monitoring and optimization with respect to dependability
In an open system of systems, loosely connected systems collaborate to provide dependable services to users. Services are enabled by the collaboration and adapted at runtime to fulfill certain dependability requirements. When systems meet at runtime, the DDIs (set up during development time) are used to check if they can provide together new services in a dependable way. If this check is passed, then collaboration can occur. To assure and maintain dependability properties, DDIs are continually executed to adapt the collaboration accordingly. Expected results shall impact the deployment and adaptation of complex systems of systems to maximize functionalities and performances, while securing dependability.
Sc3: Next-Generation Connected Dependable Functionalities
In the next generation of connected vehicles, infrastructure—and third-party software providers will be delivering situation-dependent services or additional services like apps for smartphones. In this context, DDIs will support the dependability of dynamic implementations which are based on service oriented architectures, thus enabling novel situation-dependent features, for example, in the context of autonomous and automated driving
Sc4: Dependable runtime integration for the exchange of information
In this scenario, DDIs oversee the transfer of dependability-relevant data from a service provider to a data repository in an open system where the connection between provider and repository is dynamically redefined. Data integrity and security are the prime concerns addressed here.

5 Conclusions

The physical and digital worlds are currently merging, leading to a largely connected globe. However, developments like IoT and CPS pose enormous ethical, economic and related technical questions, which we should address responsibly in the traditions of the scientific method. A key challenge that we identified and discussed here is that it is currently impossible to assure the dependability of smart CPS such as autonomous cars, swarms of drones, or networks of telehealth devices. Such systems are loosely connected and come together as temporary configurations of smaller systems which dissolve and give place to other configurations. The configurations a CPS may assume are unknown and potentially infinite. In addition, as systems connect, emergent system behaviors may arise in ways that are difficult to predict from simple superposition of the behavior of individual system elements. State-of-the-art dependability analysis techniques are currently applied during the design phase and require full a priori knowledge of system configurations. Such techniques are not directly applicable to open and dynamically reconfigured CPS.

Moving beyond the challenge, we introduced the DEIS H2020 research project, an effort focusing on the problems discussed above. The project is developing a foundation of methods and tools that lay the groundwork for assuring the dependability of CPS. In the core of this work lies the novel concept of a DDI for components and systems. DDIs are planned as an evolution of current modular dependability specifications and model aspects of the safety, reliability and security "identity" of the component. They are produced during the design phase and their profiles are stored in the "cloud" to enable checks by third parties. They are composable and executable, and facilitate dependable integration of systems into "systems of systems". The paper introduced early work in this area and discussed a wide range of cross-sectoral opportunities for industrial application. We hope to be able to report soon providing reflections on case studies and results of evaluations.

References

Armengaud E, Höller A, Kreiner C, Macher, G, Sporer H (2015) A combined safety-hazards and security-threat analysis method for automotive systems. In: Proceedings of SAFECOMP international conference on computer safety, reliability and security

Azevedo L, Parker D, Walker M, Papadopoulos Y, Esteves Araujo R (2014) Assisted assignment of automotive safety requirements. IEEE Softw 31(1):62–68

Dheedan A, Papadopoulos Y (2010) Multi-agent safety monitoring system. IFAC Proc 43(4):84–89

EC (2011) Roadmap to a single european transport area—towards a competitive and resource efficient transport system. European commission

Ferreira-Parrilla A, Kural E, Jones S (2014) Traffic light assistant system for optimized energy consumption in an electric vehicle. In: ICCVE

HRI analysis and centers for medicare and medicaid services national health expenditures (2012)

Jones S et al (2016a) V2X Based traffic light assistant for increased efficiency of hybrid & electric vehicles. In: Automotive meets electronics congress, conference paper

Jones S et al (2016b) V2X based traffic light assistant for increased efficiency of hybrid & electric vehicles, VDI Wissensforum, conference paper

PwC health research institute, new health entrants (2015) Global health's new entrants: meeting the world's consumer. https://www.pwc.com/mx/es/industrias/archivo/2015-02-global-healthcare-new-entrants.pdf

Schneider D, Trapp M (2013) Conditional safety certification of open adaptive systems. ACM Trans Auton Adapt Syst 8(2, Article 8):20

Schneider D, Trapp M, Papadopoulos Y, Armengaud E, Zeller M, Höfig K (2015) WAP: digital dependability identities. In: 26th international symposium on software reliability engineering (ISSRE'15). pp 324–329

Shift2Rail (2015) Shift2Rail strategic master plan. Shift2Rail. https://ec.europa.eu/transport/sites/transport/files/modes/rail/doc/2015-03-31-decisionn4-2015-adoption-s2r-masterplan.pdf

Structured assurance case metamodel (SACM) version 2.0, Object managment group (OMG) (2016). http://www.omg.org/spec/SACM/2.0/Beta1/PDF/

UNIFE (2014) European rail industry guide. http://www.unife.org/index.php?option=com_attachments&task=download&id=499

Windley P (2005) Digital identity. O'Reilly Media

Part IV
Safety and Testing

Smart Features Integrated for Prognostics Health Management Assure the Functional Safety of the Electronics Systems at the High Level Required in Fully Automated Vehicles

Sven Rzepka and Przemyslaw J. Gromala

Abstract The current developments in automotive industry toward automated driving require a massive increase in functionality, number, and complexity of the electronic systems. At the same time, the functional safety of those electronic systems must be improved beyond the high requirements applied today already. Designing the systems for a guaranteed lifetime on statistical average will no longer suffice. Therefore, new methods in the design and reliability assessment toward maintainable or replaceable systems are required. Prognostics and health management (PHM) provides the way for this upgrade in reliability methodology. The paper introduces a multi-level PHM strategy based on smart sensors and detectors integrated into the functional electronic units so that maintenance can be triggered if needed yet always well before the actual failure occurs in the individual system.

Keywords PHM · Health monitoring · Reliability · Functional safety · Automotive electronics · Automated driving · Smart sensors

1 Introduction

The rapidly growing activities toward automated driving not only triggers the development of a new generation of electronic systems for the automobile sector, but also the move to a new generation in reliability research. The best practice is currently based on accounting for the actual physics of failure (PoF) specific to the use case of the individual electronic system: motor control, smart head lights,

S. Rzepka (✉)
Micro Materials Center, Fraunhofer Institute for Electronic Nano Systems (ENAS), Technologie-Campus 3, 09126 Chemnitz, Germany
e-mail: sven.rzepka@enas.fraunhofer.de

P.J. Gromala
(AE/EDT3), Robert Bosch GmbH, Postfach 1342, 72703 Reutlingen, Germany
e-mail: PrzemyslawJakub.Gromala@de.bosch.com

C. Zachäus et al. (eds.), *Advanced Microsystems for Automotive Applications 2017*, Lecture Notes in Mobility, DOI 10.1007/978-3-319-66972-4_14

infotainment, etc. It includes the proactive 'design for reliability' (DfR) strategy, which aims at 'first time right' design solutions so that the first physical samples can be directly sent to the final qualification tests.

The rigorous implementation of this best design practice is even more required for the design of fully automated vehicles since number and complexity of the electronic systems needed for them will be much larger than in the current vehicles. Replacing time-consuming tests by validated numerical simulations, i.e., the move from physical to virtual prototyping, without any loss of trustworthiness and comprehensiveness is a necessity to keep the time to market and the development costs of the new products competitively low.

Nevertheless, this best current practice will no longer be sufficient. In this approach, the lifetime is the key criterion. It always means the time to end of life. Physically, it is determined by sample tests and statistical evaluation. Onset and degradation, e.g., crack initiation and propagation inside the components, in the solder joints, at board and system level are not in the scope of this approach. Neither does it allow estimates for a specific individual sample. The statistical evaluation provides information only about the typical behavior and its scatter for the total set, from which the samples were taken randomly.

Autonomous driving (AD) will change completely the way we utilize cars. In the near future, the car manufacturers will not sell cars anymore but "mobility service" instead. AD will change the "ownership" model of a car to "usership". As a result of the new mobility service, it is expected that the average use of the car per day will rise from today's 1.5 h/day up to 20 h/day in the near future. This will lead to completely new, not yet investigated, use case scenarios. They will impose significantly higher loads to control units and components in shorter time. Most state-of-the-art control unit designs and reliability assessment aim for maintenance-free systems. The anticipated heavy-duty use of AD will reduce the life span of electronic control units (ECU). Therefore, new methods in the design and reliability assessment toward maintainable or replaceable system are required.

Prognostic health management (PHM) provides the way for this upgrade in reliability methodology. Instead of just determining the ultimate lifetime on statistical average, it allows to quantify the remaining useful life (RUL) of the individual sample under its specific use conditions.

In the following sections, the paper first introduces to the PHM methodology and shows the directions of research to be conducted on the way to its introduction. Second, a multi-level PHM concept is proposed, which uses integrated sensors and signal processing for increasing the functional safety of electronics components and systems to the level required by fully automated vehicles. Finally, examples of deformation and stress analysis are provided for illustrating how changes in these parameters can be used as early warning indicators and as basis for RUL estimation models.

2 Prognostics Health Management

The objective of PHM is the detection of onset and development of anomalies and defects well before the functional performance degrades substantially and to estimate when it would leave the acceptable range, i.e., the remaining useful life of the system. This early detection shall be used for triggering dedicated maintenance activities so that the anticipated failure can effectively be prevented. The strategy of PHM is based on the monitoring of essential properties and of the service conditions relevant to the failure modes affecting the safety critical components, modules, and systems during their operation in the field. In the field of automotive electronics, the PHM strategy has not yet been implemented widely. In other application fields, PHM is already in use. For example, Rolls Royce uses Engine Health Management (EHM) to track the health of thousands of engines operating worldwide, using onboard sensors and live satellite feeds (https://www.worldfinance.com/markets/rolls-royce-is-driving-the-progress-of-the-business-aviation-market). The EHM signature will typically highlight a change in an engine characteristic. Expert knowledge is then used to turn this diagnostics signature into a prognosis. For the typical signatures, the physical root causes are known and the urgency of maintenance can be estimated.

Also for electronic systems, PHM techniques and components have already been developed (Lall et al. 2006, 2013; Wang et al. 2011). Currently, they are applied to domains like military and avionics (Vichare et al. 2004). In general, they will certainly become mandatory for future cars especially in the context of automated driving. They are able to assure the functional safety at the required high level. However, the PHM measures implemented in automotive applications will most likely be specific. In contrast to domains like avionics and military, automotive is providing solutions to a mass market. Hence, strategies like the 2-of-3 redundancy at the system level cannot be used. Of course, they are very safe in determining whether one system has failed and which one it is during any mode of system operation. The failed system simply behaves differently to the majority of the other two. However, this strategy collides massively with the rigid cost, weight, energy and form-factor constraints to be in place for automotive products. In addition, a pure 2-of-3 redundancy does not allow the estimation for how long the two systems will still work properly after the first one failed. Hence, the urgency of the maintenance cannot be quantified. Therefore, alternative strategies shall be assessed and developed specifically for the needs of automotive electronics.

In general, the PHM methods are based on pre-indicators, which allow capturing the onset and the progression of the defects. They can be derived from the physics of failure (PoF) or from data-driven (DD) approaches. The PoF schemes often involve numerical modeling and simulation to replicate and to study the physical failure mechanisms in detail. The DD approach relies on statistical methods to deduce the RUL from the actual trends of functional or assessment parameters. The fusion prognostic methodology combines PoF and DD approaches (Vichare et al. 2004) and the advantages of both methods. It reduces the uncertainty in the damage prediction by involving the full knowledge about the failure mechanisms but also taking into account the specifics of the actual system and its possible anomalies.

In order to develop the specific PHM measures for the future automotive applications, a metro map type of plan has been established (Fig. 1). On the way to the five destinations of methodology research:

- Developing the required infrastructure, sensors, electronics hardware
- Studying and characterizing the failure mode and mechanism effect by thorough analyzes (Failure Mode, Mechanism, and Effect Analysis) for both, PoF and DD, approaches
- Providing appropriate solutions to the data acquisition, management, and secure data transfer
- Performing the data fusion for reaching at one integrating global health assessment, diagnostics, and prognosis score per application

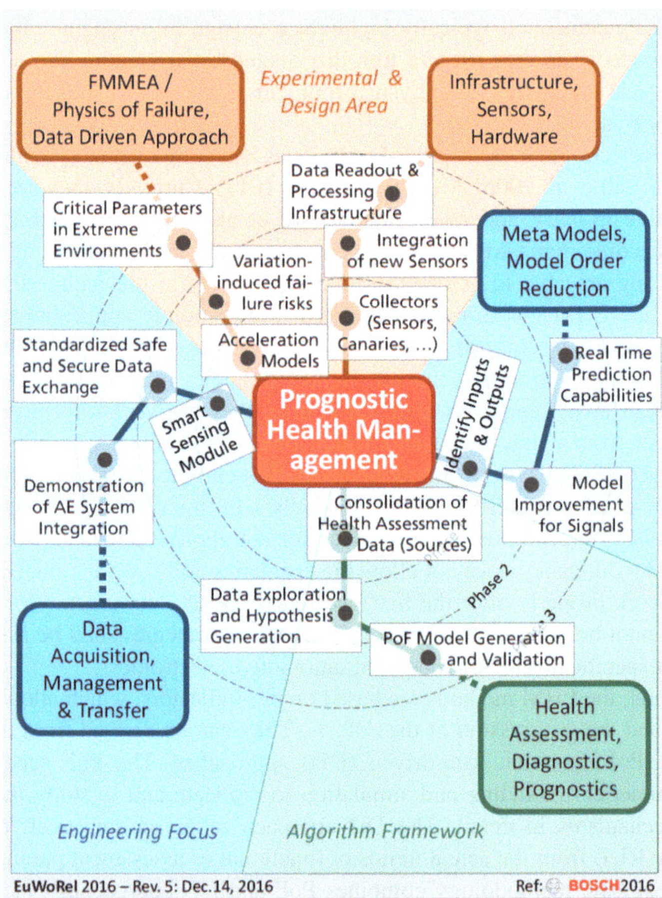

Fig. 1 PHM methodology metro map

- Establishing highly efficient yet precise metamodeling and model order reduction schemes that can be executed in each of the individual cars locally assisted by self-learning capabilities provided by cloud service

Dedicated stops have been identified and assigned to the methodology research phases 1–3, which aim at a seamless integration of the PHM strategy in the architecture of the electronics system for the future fully autonomous vehicles.

3 PHM Strategy

In order to implement fusion PHM methodology for improving the functional safety of the electronics systems as required for AD cars most efficiently, a multi-level approach is recommended (Fig. 2). Applying regular fabrication processes, highly miniaturized sensor elements are added at the lowest level that allows detecting the respective failure modes. In addition, sensor elements available in the system anyway shall measure temperature, moisture, acceleration and other environmental parameters at component and board level. Furthermore, data collection, data fusion, and signal processing capabilities dedicated to PHM need to be foreseen on the component, board, and higher system levels. These add-ons to the functional features clearly increase the cost of the system (Lall et al. 2012). However, they are substantially less expensive than the current practice of needing full-scale system redundancy, especially like in avionics. Moreover, the PHM method is able to continuously provide quantitative estimates for the RUL.

The PHM approach comprises following building blocks.

1 **Circuitry Level**: Monitoring features are added to the circuitry for watching essential device functions and for capturing the internal sensor signals. They

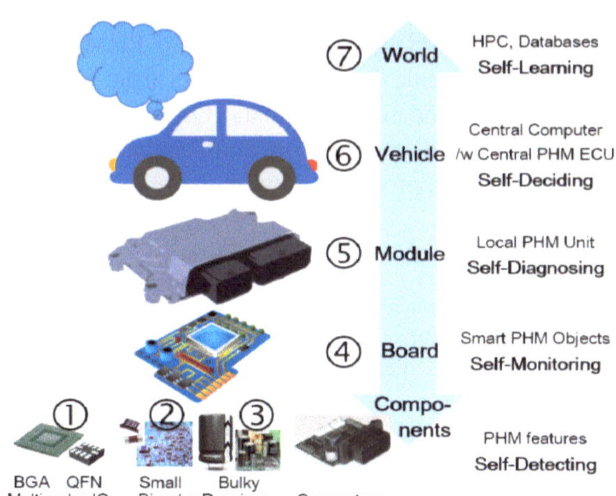

Fig. 2 Multi-level approach to implement PHM in automotive electronics

trigger the reporting of out-of-spec behaviors to the local PHM unit and the adoption of a safe mode of operation according to the new health status.

Wafer Level: Sensor features are integrated into the front-end structures for measuring functional parameters like current and voltage, operational parameters like temperature, humidity, and acceleration, but also integrity parameters like mechanical strains and stresses highly sensitively at critical points inside the components. They provide damage self-detection capabilities to these devices. For example, crack initiation and growth at any interface of the mold to die, adhesive, or lead frame will change the local strain distribution and its temperature dependence (Gromala et al 2015; Schindler-Saefkow et al. 2012).

Component Level—IC: Adding a few wire bonds or micro bumps (inside) and solder joints (outside, Fig. 3) at places of highest loads provides early warning indicators to high pin-count components (Frühauf et al 2016). FEM simulation validated by experiments can quantify the RUL of the essential interconnects for the device function based on the lifetime of the PHM canary features.

2 **Components—bipolar passives**: Monitoring the health status of small canary passives attached to the printed circuit board (PCB) at high stress areas by specifically designed weak links (Chauhan et al. 2014) allows quantifying the RUL of the functionally essential components by validated simulation schemes.
3 **Large components**: Usually, large components like capacitors or connectors are glued to the board in addition to the soldering of the pins. Once the adhesive starts debonding (due to moisture, etc.), the local PCB warpage changes well before any functional failure. Strain sensors inside the PGB can capture these changes. FEM simulation and modeling can estimate the RUL as described below.
4 **Board level**: Besides the objects listed so far, dedicated PHM system in packages (SiP) comprising miniaturized sensors and microcontrollers may also be attached to the boards of safety relevant modules for characterizing the relevant environmental and service conditions. In total, this full set of PHM objects provides the data required for self-monitoring of the electronic system according to the new PHM approach.
5 **Module Level**: One of the onboard PHM objects will be upgraded to be the local PHM unit. It collects the data from all PHM objects and canary features, processes, and fuses this data. Of course, the full-size data processing by 3D FEM simulation based on multi-scale, multi-physics, and fracture or damage mechanics approaches requires large computational resources and takes a few

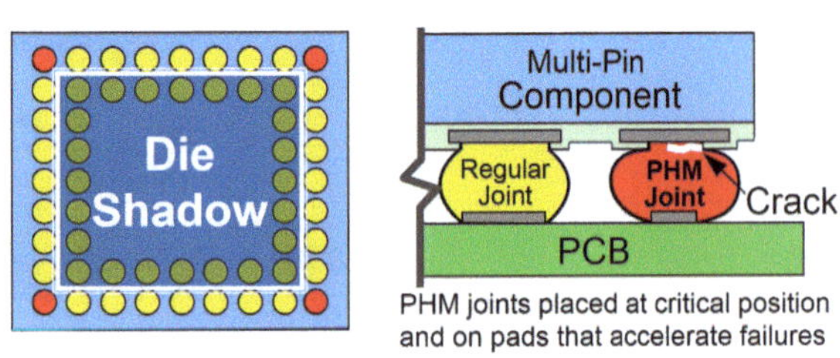

Fig. 3 PHM canary feature for multi-pin components (Frühauf et al. 2016)

hours per single case. This is by no means implementable into a local PHM unit. Therefore, a meta-modeling approach will be used for PHM monitoring and decision making on maintenance or repair in due time. Its basic concept is similar to that of the response surfaces. The PHM input data is related to the response value(s), e.g., to the RUL, by approximation models. The response values are determined outside the car and not in real time by the complex FEM simulations, which are validated by experiments. The representation of the relation between PHM input and RUL response values can be minimalistic. Polynomials or similarly simple models are sufficient. The computational resources of microcontrollers are adequate for these metamodels, which allow interpolation between nearest neighbors if the response RUL value is not yet available in the local lookup table directly. In fact, they can also fuse the individual component RUL responses deduced this way from environmental, operational, and integrity inputs into a local health score, which expresses the health status of the particular module by a single value. The result of this self-diagnostics is communicated to the central PHM ECU together with the set of input data in all cases that required interpolation.

6 **Vehicle Level**: The central PHM ECU collects the diagnostics information from all safety relevant modules (e.g., steering, motor control, battery and energy management, etc.) and deduces the global health score, which expresses the health status of all electronic systems. It can be communicated to the passenger. In parallel, it allows the AD car to self-decide on the maintenance calls as needed.

7 **Global Level**: The communication channel from the central PHM ECU to the local PHM unit with access to the PHM input data on one side and to the cloud services (database and HPC) on the other side constitutes a self-learning scheme. In addition to the interpolation by metamodels, which provides an immediate local solution but is of approximate nature, the set of PHM input data is sent via a secure link to the cloud based PHM database or—in case, the correct RUL response value is neither available there—to the high-performance computing (HPC) cluster. The exact response value (found in the database or calculated by HPC) is sent back to the vehicle and added to the lookup table of the particular PHM unit—perhaps only if significantly different to the interpolated solution. This way, the PHM system will benefit from a self-learning scheme (without overflowing), to which not just that particular car but many similar ones are able to contribute.

This pragmatic multi-level approach to PHM sketched is seen as a first draft version only. It will need to be enriched further by adopting and developing additional methods for identifying onset and propagation of potentially critical degradation mechanisms—not limited to the field of thermo-mechanical risks, which has been in the focus here. It will require truly multi-discipline research to make AD cars ready for the regular mass market.

4 PHM Indicators and Parameters for the RUL Estimation

This chapter visualizes the application of the PHM approach for failure detection inside the ECU. There are many different sensors and source of signals that can be used for the PHM purpose. A straight forward way is to utilize sensors or other components already integrated into the current ECUs but not yet used for PHM. The temperature sensors can be seen as a simple example. They are integrated into each ECU and exist even in most of the application-specific integrated circuits (ASIC). Moisture sensors and accelerometers would be further examples. They all can provide information on service and environmental conditions, based on which RUL of electronic components and systems can be estimated.

However, the features mentioned so far cannot directly characterize the loads in the electronics components. Hence, the RUL estimates rely on model assumptions regarding the relation between the conditions actually measured and the remaining life of the components. Another possibility is to use dedicated features or sensors for directly capturing the characteristic parameters of the most interesting failure modes. Here, piezoresistive silicon-based sensors allow measuring the mechanical stress on the top of the silicon die directly, which is responsible for failure modes like delamination and die fracture. Below, we will describe an example of utilizing iForce sensor for in situ detection of the delamination in automotive ECU.

The test vehicle designed for the first study contains six discrete packages (DPAK) located at three different positions on the PCB (Fig. 4). The PCB has two layers, with full layer copper at the top and the bottom. The thickness of the PCB is 1.6 mm. In order to distribute the heat from the DPAK during active operation, 25 thermal vias are made in the PCB.

On the back side of the PCB (Fig. 4b), there are three IForce piezoresistive stress sensors. The stress sensors are to be used for feature extraction, based on which the state of health of the control unit is estimated. The entire test vehicle was

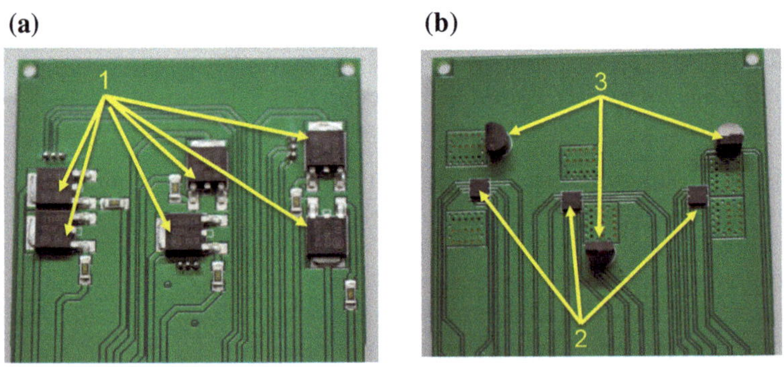

Fig. 4 Test vehicle: **a** *top side view*, **b** *back side view*; 1—DPAK, 2—stress sensor, 3—temperature sensor

overmolded using standard molding technology. Our focus was on delamination between molding compound and PCB on the side, where DPAK components were located. In order to observe the growth of delamination, some of our samples were pre-delaminated by special surface treatment.

The sensor used in the study is an iForce, piezoresistive silicon-based stress sensor. The sensing elements are created by the channels of MOSFET transistors in a current mirror circuit as shown in Fig. 5. Its specific construction enables the stress measurements with high spatial resolution. The current mirror circuit is very sensitive to differences in parameters of the transistors. The channels of MOSFETs are oriented in such a way that the change in stress is changing their resistivity. Both of these properties are used to measure the stresses with very high sensitivity. By measuring the currents flowing through both branches of the current mirror, one can calculate an in-plane shear stress, σ_{XY} (1), and the difference in in-plane normal stress components, $\sigma_{xx} - \sigma_{yy}$, from the following relationships (2):

$$\sigma_{xy} = \frac{1}{\pi_{11}^{(n)} - \pi_{12}^{(n)}} \frac{I_{out} - I_{in}}{I_{out} + I_{in}} \tag{1}$$

$$\sigma_{xx} - \sigma_{yy} = \frac{1}{\pi_{44}^{(p)}} \frac{I_{out} - I_{in}}{I_{out} + I_{in}} \tag{2}$$

where π_{11}, π_{12}, π_{44} are the piezoresistive coefficients of N and P doped silicon in the in-plane and the shear directions, respectively, and I_{IN}, I_{OUT} are the currents measured at the input and at the output of current mirror, respectively.

Each sensor consists of a matrix of sensing cells. The sensor with 12 sensing cells in a 4 × 4 array is used. It is to be noted that four corner cells are inactive as shown in Fig. 5. The sensor is packaged in a standard microelectronics land grid array (LGA) package. The silicon die is first attached to an interposer using a thermally conductive adhesive. Electrical connections are formed by wire bonds. Additionally, a dummy ceramic component is soldered on the interposer. The whole

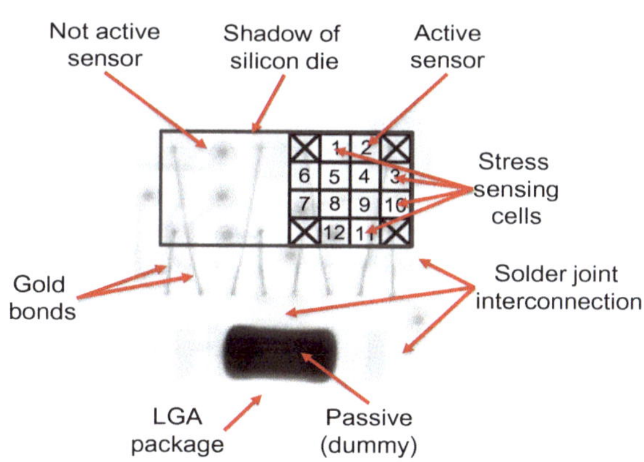

Fig. 5 Silicon based, piezoresistive stress sensor

construction is overmolded using an epoxy molding compound. The final dimension of the package is 3 mm × 3 mm × 1 mm. The LGA package is attached to the PCB through the standard solder joint interconnection.

Finding the failure mechanism and reliability models to quantitatively evaluate the susceptibility to failure requires understanding the relationship between requirements of the product and physical characteristics. FMMEA relates the interaction between materials and stresses during operational condition. Figure 6 shows schematically the FMMEA process.

An application of the algorithm for fault detection and classification is presented in the studies (Palczynska et al. 2016; Prisacaru et al. 2017). The procedure employed requires the state of health. It is obtained by using the Mahalanobis Distance (MD) as health criterion. The MD quantifies the distance between an individual value such as the difference stress of one component and the center of its distribution. If the distance is larger than the specified fault threshold the respective component is considered as failed. Figure 7 presents an example. MD is used to detect the delamination between PCB and compound. In Fig. 7a, the difference stress is plotted at the end of each thermal cycle. The 'healthy package'—without any pre-delamination—is shown in the upper part. It exhibits a continuous behavior with very low scatter and just a shallow drift. In the lower part, the 'failed package' is shown, which has an intentional pre-delamination right from the beginning of cycling test. A sudden change in the difference stress occurs after 30 cycles. The delamination that propagated along the top side of PCB, where the DPAK components are soldered, had obviously reached the site of the sensor mounted at the lower side of the PCB in the center of the DPAK (Fig. 4). Figure 7b shows the results of MD evaluation for both cases. On the top, we see the results for the package without defect. All the points are below both, warning and failure, thresholds. No anomaly is detected. In contrast to this, the MD values exceed the

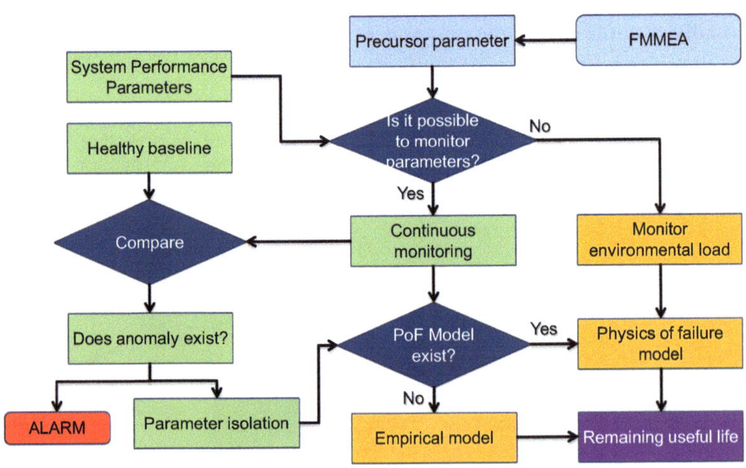

Fig. 6 FMMEA methodology adapted for current study (Gromala et al. 2015)

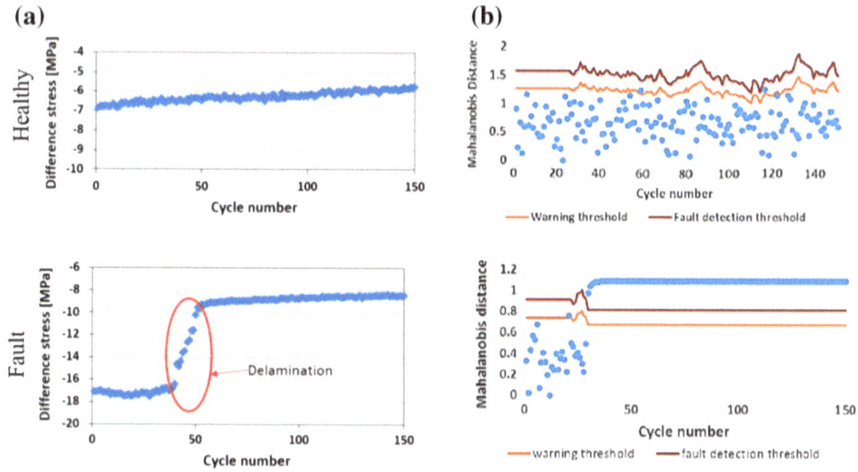

Fig. 7 Health monitoring during thermal cycles: **a** Stress evolution of a healthy (*top*)/a damaged (*bottom*) package, **b** Anomaly detection prior to the failure

limits after 30 cycles in the results of the package with the pre-delamination. The change in the stress state surely triggers an alarm. The MD criterion can serve as early warning indicator.

In order to recognize what kind of failure happened, MD is not sufficient. Another algorithm must be used, such as support vector machine (SVM). SVM is a supervised learning model that analyzes the data and allows for failure classification. Based on that technique, we can clearly identify what kind of failure happens and estimate its severity. Subsequently, both local and global health score can be obtained and the RUL can be estimated.

Acknowledgements The authors would like to thank the PHM team of EuWoRel 2016 for the fruitful discussion on the PHM metro map. In particular, we thank the track owners: D. Vanderstraeten (OnSemiconductor), J. Arwidson (Saab), E. Tsiporkova (Sirris), S. Kunath (Dynardo). We are looking forward to the work in 'smartSTAR', the PENTA project supported by BMBF (Germany) and VLAIO (Belgium).

References

https://www.worldfinance.com/markets/rolls-royce-is-driving-the-progress-of-the-business-aviation-market

Chauhan P, Mathew S, Osterman M, Pecht M (2014) In situ interconnect failure prediction using canaries. IEEE Trans Device Mater Reliab 14(3):826–832

Frühauf P, Gromala P, Rzepka S, Wilke K (2016) Electronic component comprising a plurality of contact structures and method for monitoring contact structures of an electronic component. Patent application EP3086132A1

Gromala P, Palczynska A, Han B (2015) Prognostic approaches for the wirebond failure prediction in power semiconductors: a case study using DPAK package. In: 16th international conference on electronic packaging technology, ICEPT. pp 413–8

Lall P, Islam MN, Rahim MK, Suhling JC (2006) Prognostics and health management of electronic packaging. IEEE Trans CPMT 29(3):666–677

Lall P, Lowe R, Goebel K (2012) Cost assessment for implementation of embedded prognostic health management for electronic systems; ASME 2012. In: international mechanical engineering congress and exposition, IMECE 2012, vol 9. pp 775–85 (Issue Parts A and B). doi:10.1115/IMECE2012-93058

Lall P, Lowe R, Goebel K (2013) Comparison of prognostic health management algorithms for assessment of electronic interconnect reliability; ASME 2013. In: Proceedings of International technical conference and exhibition on packaging and integration of electronic and photonic microsystems, InterPACK 2013, vol 1. doi:10.1115/IPACK2013-73252

Palczynska A, Prisacaru A, Gromala P, Han B, Mayer D, Melz T (2016). Towards prognostics and health monitoring: the potential of fault detection by piezoresistive silicon stress sensor. In: 17th international conference on thermal, mechanical and multi-physics simulation and experiments in microelectronics and microsystems, montpellier. doi:10.1109/EuroSimE.2016.7463344

Prisacaru A, Palczynska A, Gromala P, Han B, Zhang GQ (2017) Condition monitoring algorithm for piezoresistive silicon-based stress sensor data obtained from electronic control units. In: 67th IEEE electronic components and technology conference. Orlando

Schindler-Saefkow F, Rost F, Faust W, Wunderle B, Michel B, Rzepka S (2012) Stress chip measurements of the internal package stress for process characterization and health monitoring. In: 13th International conference on thermal, mechanical and multi-physics simulation and experiments in microelectronics and Microsystems. Cascais, Portugal. doi:10.1109/ESimE.2012.6191746

Vichare N, Rodgers P, Eveloy V, Pecht MG (2004) In situ temperature measurement of a notebook computer—a case study in health and usage monitoring of electronics. IEEE Trans Device Mater Reliab 4(4):3–658

Wang Y, Miao Q, Pecht M (2011) Health monitoring of hard disk drive based on Mahalanobis distance. In: Proceedings of prognostics and system health management conference, vol 1. Shenzhen, pp 1–8

Challenges for the Validation and Testing of Automated Driving Functions

Halil Beglerovic, Steffen Metzner and Martin Horn

Abstract In this paper, we will explore challenges for the validation and testing of Automated Driving Functions (ADF), which represent one of the major roadblocks for successful integration of emerging technologies into commercial vehicles. We provide an overview of current methodologies used for validation and testing, focusing on the missing parts. Furthermore, we give an insight into promising methodologies, frameworks, and research areas which aim to reduce current testing and validation efforts.

Keywords Challenges · Validation · Testing · Automated Driving Functions (ADF) · Advanced Driving Assistance Systems (ADAS)

1 Introduction

In order to satisfy the increasing demand for safety, reliability, and comfort of commercial vehicles, manufacturers and research groups put great effort in the development of new and sophisticated driving functionalities throughout the decades. This development has undergone several phases and it started with the driver assistance systems which required constant driver control, e.g., Cruise Control (CC), or were active only in certain situations, e.g., Emergency Brake Assist (EBA), Electronic Stability Program (ESP).

H. Beglerovic (✉) · S. Metzner
AVL List GmbH, Hans List Platz 1, 8020 Graz, Austria
e-mail: halil.beglerovic@avl.com

S. Metzner
e-mail: steffen.metzner@avl.com

M. Horn
Institute of Automation and Control, Graz University of Technology, Inffeldgasse 21B/I, 8010 Graz, Austria
e-mail: martin.horn@tugraz.at

C. Zachäus et al. (eds.), *Advanced Microsystems for Automotive Applications 2017*, Lecture Notes in Mobility, DOI 10.1007/978-3-319-66972-4_15

With the increase in the processing power of computers, researchers and OEMs were able to include more sophisticated algorithms and sensors into the vehicle, what led to the development of Advanced Driving Assistance Systems (ADAS). The new technologies allowed the development of Adaptive Cruise Control (ACC), Lane Keep Assist (LKA), Lane Change Assist (LCA), traffic signs, pedestrian and vehicle detection, etc. Even though these systems are more complex and sophisticated, the driver still needs to monitor them continuously in order to compensate unforeseen conditions on the road or misbehavior of the algorithm. Nevertheless, the safety benefits of ADAS systems have led governments to make laws which obliges manufactures to include some of them into commercial vehicles (e.g., ABS and ESP), showing that these systems have become an irreplaceable part of the driving experience.

In the recent years, a significant rise in the set of functionalities of Automated Driving Functions (ADF) could be observed, as they are increasingly introduced into vehicles. ADF represent the next technological step as they aim to provide a driving experience where constant monitoring of the system is not needed. In its core, ADF combine several ADAS functions in a comprehensive and complex system. For example, a Highway Pilot ADF is combining ACC, LKA, LCA, traffic sign and vehicle recognition in order to successfully drive the vehicle on a highway. Many manufactures like Tesla, Daimler, Otto, etc., have already demonstrated various capabilities of highway pilots.

Current technology allows limited automated driving on specific scenarios and the aim for the future, as shown in Fig. 1, is to enable full autonomy where human interaction is not needed at all. These systems should be able to perceive and understand the environment in order to act accordingly to all possible situations encountered in real traffic. Such systems would open the door to new types of transportation where fleet of automated vehicles could be shared between people, effectively lowering the overall number of used vehicles. In addition, the mobility of elderly people or people with disabilities would increase drastically.

The main barrier for the commercialization of ADF is the need to test them on a theoretically infinite variety of scenarios, including a huge number of parameters to be varied to reflecting the reality. In order to keep up with the high commercialization demand, rapid developments and advancements in technology, testing, and

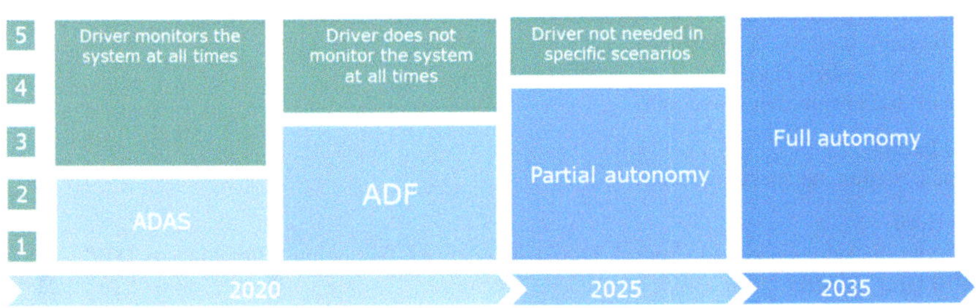

Fig. 1 Development of assistance systems

validation procedures need to lower the current testing effort, and in addition, new innovative methods need to be developed. Watzenig and Horn (2016) state that there is an increasing demand for methodologies and validation procedures that will enable the development of ADF which allow greater safety, traffic flow optimization, reduced emission and enhanced mobility.

2 Challenges for Validation and Testing

The challenges for validation of ADF are encountered on both sides, methods and tools. On the method side, the safety of the intended function (SOTIF) needs to be ensured while remaining economically feasible. On the other hand, the tools side needs to deal with new signal types and resolve the problem of transferring huge amount of data between the control units, which is out of the scope for the classic transfer protocols used in today's vehicles.

2.1 Complexity of Automated Driving Functions

A general structure of an ADF can be seen in Fig. 2. The task of an ADF is to perceive and understand the environment and take adequate actions depending on the current situation. In order to accomplish this task, ADF use an increasing number of heterogeneous sensor systems and complex algorithms fusing and interpreting the data of the dynamic environment. This sensor inputs are combined with existing knowledge coming either from maps, vehicle-to-vehicle, or vehicle to infrastructure communication. All of these inputs, together with the internal states of the vehicle, e.g., velocity, engine speed, etc., are needed by the decision-making algorithms in order to predict and plan adequate trajectories.

However, the inputs coming from the heterogeneous sensor introduce new data types, e.g., object lists, images, radar data, and current software component verification tools are not able to handle them accordingly. In addition, the programming environments used for development also need to provide an interface to these new

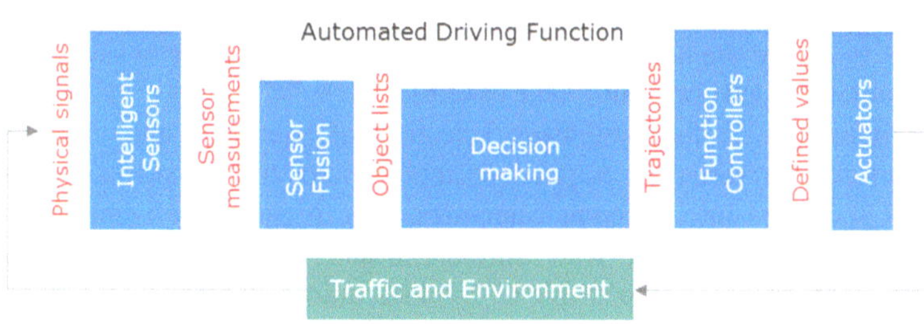

Fig. 2 Structure of an Automated Driving Function

data types. Furthermore, with the advancement of technologies like deep neural networks, dedicated hardware is introduced and developers need to be able to interface both the models and hardware into their existing development environment. Interfacing all these data streams with standard tools during the development phase is still an open issue, as there are still no communication standards.

2.2 Variation of Scenarios and Parameters

Winner et al. (2013) and Wachenfeld et al. (2015) state that an ADF should be driven for more than 240 million kilometers without fatalities to prove that they are, at least, not performing worse than human drivers. Performing such an extensive validation, using classical validation methods on public roads or proving grounds, would not be possible in a sustainable frame. If we take into consideration a typical development cycle, with several iterations of testing and validation, it becomes clear that such extensive test runs are not economically feasible.

On the other hand, simulation has been proven to be a very promising tool for successful development and testing of ADF. The most important advantage is the ability to easily vary the various scenarios and parameters, ensure reproducibility and, in the case of pure software in the loop (SiL), even run test cases in parallel and faster than real-time.

However, even if simulation is used for the scenario and parameter variation, the ADF could theoretically be tested on 10^{12} different test cases within various scenarios, traffic, and environment parameter variations. This type of full factorial testing is not feasible and new advanced methods are needed for appropriate scenario selection and parameter variation.

2.3 Scenario Selection and Test Generation

In order to successfully validate an ADF, an appropriate selection of scenario-subsets from all possible scenario variations, which will lead to a sufficient scenario coverage, is needed. Only a small portion of all available scenarios presents a challenge for the ADF and could potentially lead to faulty behavior. It is important to reduce the number of considered scenarios in order to save time, costs and resources needed for testing and validation.

The main task is the systematic selection of scenarios and corresponding parameters for the variation. A possible approach for the scenario selection is to group all scenarios into different categories, e.g., highway, left turn, right turn, country road, etc., together with different traffic behavior. With this type of grouping methodology it is possible to conduct validation on certain test runs and exclude test runs with similar characteristics from further consideration.

An opposite approach could be applied by identifying those scenarios leading to the most critical behavior of the ADF. Then by knowing these most critical scenarios, which can be taken from existing accident databases, we could expand the test to similar runs and conclude where the critical area stops.

Additionally, a subset of scenarios could be selected depending on specific regulations, imposed by the laws from different countries.

After appropriate scenario selection, an automated test generation procedure is needed. This procedure should be able to generate appropriate test cases taking into consideration the capabilities and level of autonomy of the ADF. Furthermore, no unified metrics, references nor testing criteria exist up to today. The test generation is usually carried out by engineers tailored to meet their specific needs. Better objective evaluation methods are needed which can derive, with some certainty, the overall behavior of the model from just a subset of test cases.

3 Current Methodologies/Technology Overview

Today's ADF functions are developed using environment simulation systems like VTD from Vires, PreScan from Tass or IPG CarMaker (there are much more available). Especially in the beginning of the development process, for both the algorithm and software development, those simulation systems are well accepted. All available environment simulation systems have strengths and weaknesses depending on the origin domain (vehicle dynamics, driver simulator, etc.) and allow manual design and simulation of a wide range of scenarios. Dependent on the complexity of the respective scenario, setup and design of those take a lot of time.

As for usual control units, the functionality and the electronics are integrated and tested on HiL systems. The HiL systems are either used for testing one single control unit or testing several units in compound on an integration HiL. Usually, setup and maintenance need a lot of effort. Variability and the possibility to influence the inputs of the control unit independent from the environment of the vehicle overcome these disadvantages.

After the integration of the functionality into the vehicle, testing and validation is nowadays executed on test tracks and public roads (Benmimoun 2017). Especially on public road, the reproducibility of scenarios is very difficult and can be very time consuming. Although simulation is used for algorithm development extensively, up to now the trust in this methodology is not high enough for testing with prototype vehicles.

In some cases, already ground truth measurement systems are used (Fritsch et al. 2013). Such systems are able to collect the data of the environment with higher precision than the mass productions sensors inside the vehicle. These systems enable the engineers to compare the sensor output of both vehicle sensor and ground truth measurement system, and facilitate high improvements in debugging.

However, reproducing such measured scenarios in simulation environment to debug the source code in detail still means a lot of manual work. In some cases, limited tool support is available, which converts such measurements in reproducible scenarios to be executed in environment simulation tools.

Forthcoming autonomous driving systems are further extending the testing and simulation requirements of the safety and domain control functions. The vehicle needs to be able to anticipate road hazards in advance, and adjust driving strategies accordingly to increase driver's trust. Thus, there needs to be a way to provide the vehicle with an awareness of the road environment beyond the reach of its onboard sensors.

4 Validation—Global Approach

It is still unclear, how validation, homologation, and certification of automated driving functions above SAE (www.sae.org) automation level 2 shall be done. However, several research projects on national and international level are taking care about this topic. There are few points nearly all experts do agree on:

- The solution uses simulation for a major part of scenario-execution;
- A clever combination of methods and validation environments (SiL, HiL, test-track, public road, etc.) is necessary;
- The number of test cases/scenarios will still be quite big.

One of the research projects dealing with testing and validation optimization is the European ESCEL research project ENABLE-S3 (www.enable-s3.eu). The aim of the project is to reduce current efforts needed to test and validate an automated system by 50%. The project is divided into two comprehensive parts as shown in Fig. 3. First, the validation methodology deals with novel approaches for scenarios and metrics selection, data acquisition and storage, and test optimization. In order to run the generated test cases, the second part of the project focuses on a reusable validation framework which can support seamlessly various development stages (MIL, HIL, VIL, Proving ground, etc.).

5 Supporting Tools in the Validation Task

The environment simulation tool is the central tool on all development and validation levels involving simulation. Such environment simulation systems are already available from several suppliers and usually include sensor models that can be directly connected to the ADAS/ADF controllers. Dependent on the accuracy of the models, a huge amount of processing power may be required. This includes models involving ray tracing methods necessary for detailed camera or radar models. However, for a lot of cases this accuracy is not necessary or even obstructive. If very accurate sensor models are used and the robustness regarding

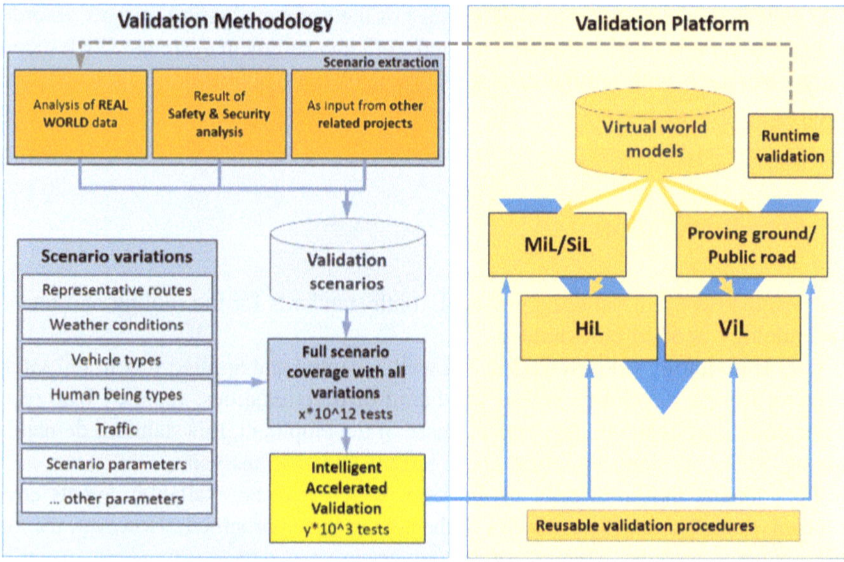

Fig. 3 Modular validation and verification framework of the ENABLE-S3 project (www.enable-s3.eu)

special effects of the sensor shall be tested, it is also necessary to model the environment in the same level of detail to trigger these effects. In such cases, a phenomenological model may be a better choice as the modeling of the scenario is made much easier. This is also valid for other parts of the model, e.g., the vehicle model may be reduced in details, if the test case does not contain stability critical maneuvers. Therefore, simulation platforms need to support the exchange of models dependent on the parameterization of the test case.

Additionally, mixed/real development and validation environments will be part of the validation as they are also in today's environment. But to enable these new methods of stimulation (e.g., for radar and camera), new data types (object lists and raw/streaming data) are necessary to be used, e.g., vehicle in the loop is a well-established environment for the development of powertrain-related functions. This environment could also be used for ADAS/ADF validation. This raises the complexity of the vehicle tests in the same way as the complexity of the vehicle.

Even if methods are found to reduce the number of test cases, there are still millions of kilometers which need to be driven either virtually or on real roads. In addition, test cases need to be handled in a structured manner and if no real hardware is involved it is possible to execute these test cases in parallel on the cloud. This coordinated distribution of the test cases and collection of the results is also an important requirement to the simulation platform as well.

However, it will not be possible to execute all test cases virtually or with only partial hardware. Some of the test cases or scenarios still need to be executed in real environment or test tracks. The data collected should be as close as possible to the data collected in simulation to enable usage of the same evaluation metrics. In order to accomplish this metrics transfer, ground truth measurement systems which record the environment with similar sensor as the vehicle but with higher precision, are necessary.

6 Standardization

For development and validation of ADF, both standards for the tooling and for the methodology need to be extended.

On methodology side ISO26262 is a well-accepted and applied functional safety standard for the development and validation of functionalities, realized with electronic control systems. In the concept phase of development, this standard demands, among others, an item definition and a hazard and risk analysis.

Both require the knowledge of the item (which may be available) and its environment, as well as the dependencies of the item to the environment. However, the fact that scenarios and the variation of the environment parameters (daylight, weather, temperature, etc.) are infinite restricts the direct application of ISO26262 on automated functions with automation level 3 and above. However, there needs to be a definition of a finite testing space regardless which method and which tools will be necessary for validation. Finding this testing space is even harder considering that all implementations of the ADF functionalities will have different strengths and weaknesses. A good example is the development of the vehicle dynamics functionality. In all test cases defined by authorities or car-magazines, for e.g., brake distance or stability during a lane change maneuver, the systems from different manufacturers react in a similar manner. But comparing those systems outside of this test space leads to significant differences as the developers are focusing on the defined test cases.

Therefore, the test case space for autonomous driving functionalities needs to be chosen wisely to avoid over- and under-optimized areas. In the worst case, this may lead to a product passing all tests but being still not useful for the customers.

ISO26262 has been published at a time where electronic control units have been already used for many years also for safety critical applications. Hence, it contains a lot of experience gained through the development of those control units. The same is to be expected for publication of the standard covering automated driving.

On the tools side the same problems as on control unit site occur as all bus systems, especially the CAN-bus, are optimized for signal-based controls-communication. This kind of communication will still play a big role in the future, but with new functions considering the objects of the environment, also object-based communication becomes necessary. For small objects containing few data, the CAN-bus may still be used, although the CAN-drivers need to be adapted. But the CAN-communication description and well known DBC-file format cannot be applied. Here it is necessary to establish new formats describing the exchanged objects and bus systems able to handle this kind of communication properly. This adaption needs to be applied to simulation, measurement and evaluation tools and on all related file types.

7 Conclusion

Currently, the main approaches for validating the ADF are based on proving ground or public road test drives. The challenges that arise from testing on proving ground or public road are that they are very expensive, time consuming, requiring huge effort and are hard to reproduce. Based on these challenges, simulation offers a solution as we are able to achieve high reproducibility with low effort. However, the question of how to select proper scenarios and parameter variations which will cover the infinite set of variation in a comprehensive manner is still an open issue. New methodologies and frameworks, able to automatically generate adequate test cases from input requirements or use ground truth data are necessary. In addition, standards dealing with new data types and new communication protocols need to be established. Furthermore, unified metrics, references and testing are needed for the overall standardization of testing procedures for ADF.

Acknowledgements The project leading to this application has received funding from the European Union's Horizon 2020 research and innovation programme under the Marie Skodowska-Curie grant agreement No. 675999.

References

Benmimoun M (2017) Effective evaluation of automated driving systems. SAE technical paper, 28 Mar 2017

Fritsch J, Kuhnl T, Geiger A (2013) A new performance measure and evaluation benchmark for road detection algorithms. Intelligent Transportation Syst (ITSC)

Wachenfeld W, Winner H (2015) Die freigabe des autonomen fahrens. In: autonomes fahren. Springer, Berlin, Heidelberg

Watzenig D, Horn M (eds) (2016) Automated driving: safer and more efficient future driving. Springer

Winner H, Wachenfeld W (2013) Absicherung automatischen Fahrens, 6. FAS-Tagung München, Munich

www.sae.org—society of automotive engineers

www.enable-s3.eu—European initiative to enable validation for highly auto-mated safe and secure systems

Automated Assessment and Evaluation of Digital Test Drives

Stefan Otten, Johannes Bach, Christoph Wohlfahrt, Christian King, Jan Lier, Hermann Schmid, Stefan Schmerler and Eric Sax

Abstract Within the last decade, several innovations in the automotive domain were introduced in the field of driver-assistance systems (DAS). As technology rapidly advances toward automated driving this trend further continues, integrating more intelligent, interconnected, and complex functionality. This results in a constantly expanding space of system states that need to be validated and verified. Approaches for virtualization of test drives such as X-in-the-loop (XiL) are in focus of current research and development. In this contribution, we introduce a concept for automated quality assessment within a randomized digital test drive. Our aim is to analyze and assess the continuous behavior of automotive systems during multiple realistic traffic scenarios within a simulated environment. Therefore an analysis and comparison of current test approaches and their verification and validation goals are conducted. These results are utilized to derive requirements and constraints for an automated assessment. In comparison to established systematic test

S. Otten (✉) · J. Bach · C. King · E. Sax
FZI Research Center for Information Technology, Haid-Und-Neu-Str. 10-14, 76131 Karlsruhe, Germany
e-mail: otten@fzi.de

J. Bach
e-mail: bach@fzi.de

C. King
e-mail: king@fzi.de

E. Sax
e-mail: sax@fzi.de

C. Wohlfahrt · J. Lier · H. Schmid · S. Schmerler
Daimler AG, 71059 Sindelfingen, Germany
e-mail: christoph.wohlfahrt@daimler.com

J. Lier
e-mail: jan.lier@daimler.com

H. Schmid
e-mail: hermann.s.schmid@daimler.com

S. Schmerler
e-mail: stefan.schmerler@daimler.com

C. Zachäus et al. (eds.), *Advanced Microsystems for Automotive Applications 2017*, Lecture Notes in Mobility, DOI 10.1007/978-3-319-66972-4_16

approaches, our concept based on a continuous assessment of the entire test drive constituting multiple driving scenarios. To consider the continuous behavior and parallel assessment of different functionality, a distinction between activation conditions and test conditions is conducted. Additionally, the hierarchization of conditions allows identification and evaluation on different abstraction levels. We include general assessments in addition to system and function-specific behavioral assessments. The approach is elaborated on an example use case of an Adaptive Cruise Control (ACC) system.

Keywords Automated driving · ADAS · Automotive testing · Verification and validation · Hardware-in-the-Loop · Digital test drive · Scenarios · Assessment and evaluation

1 Introduction

The development of systems and algorithms for automated driving steadily evolve from prototypical realization to productive use (Bengler et al. 2014). Innovations in sensor systems, high precision maps, electronic horizon, and online services lead to an increased sphere of environment perception in future vehicles (Bach et al. 2017). These improvements result in new degrees of freedom for assisting and automating features and pose demands for innovative test and validation methods. Furthermore, the space of variants required for sufficient test coverage expands with increasing functional cross-linkage beyond the established vehicle domains. To constantly ensure adequate quality and safety of these novel features, virtualization approaches are emphasized due to the required test kilometers in real world (Helmer et al. 2015). The digital test drives result in a demand for new methods for evaluation of automotive systems regarding the parametrization, interpretation, and assessment of complex driving and traffic scenarios.

Established test concepts focus on systematical testing based on test catalogs (Sax 2008). It is important to emphasize that the test cases defined in these catalogs are developed synthetically to ensure coverage of requirements based on experience and knowledge of specialized test engineers. With highly automated driving, established specification-based tests are no longer sufficient to cover the increasing variant space of possible scenarios and situations. These systematic tests represent relevant corner points that prove the correct functionality for highly relevant individual situations. To cover the regions between these distinct corner cases, statistical methods for derivation of driving scenarios will be in focus. As a result, the state of the simulation at a certain time is not known a priori and comparison of simulation results for different parametrizations of the test object requires complex assessment. A simple pass/fail statement is not always possible for evaluation of complex driving scenarios. Consequently, we can see that the established testing methods are no longer sufficient for testing of highly automated systems (Helmer et al. 2015). To tackle this challenge, we present an approach for assessment and evaluation of digital test drives based on a Hardware-in-the-loop (HiL) system.

The paper is structured as followed: Sect. 2 presents state of the art regarding systems engineering and testing in the automotive domain. The derived requirements and constraints for the assessment of digital test drives are presented in Sect. 3, followed by our concept for evaluation in Sect. 4. The concept is applied and discussed on an exemplarily ACC system in Sect. 5. Conclusion and outlook for future work are presented in Sect. 6.

2 State of the Art in Automotive Testing

2.1 Test Processes and Methodologies

Established development processes for embedded systems in the automotive domain follow the V-Model (Sax 2008). It divides the development process into several steps, whereas the left part covers all design activities, the right part comprises verification and validation (Weber 2009). Automotive SPICE (2015) provides a guideline for implementation of the V-Model and therefore serves as a process reference. Figure 1 gives an overview of the activities based on the V-Model as well as associated test technologies.

Several methods for testing are established during the different development phases. The test activities can be separated into real-world testing and simulation-based approaches. Real-world testing is achieved in context of rapid prototyping within early development stages as well as field trials in later development stages. Real-world tests follow semiformal textual test catalogs and remain indispensable

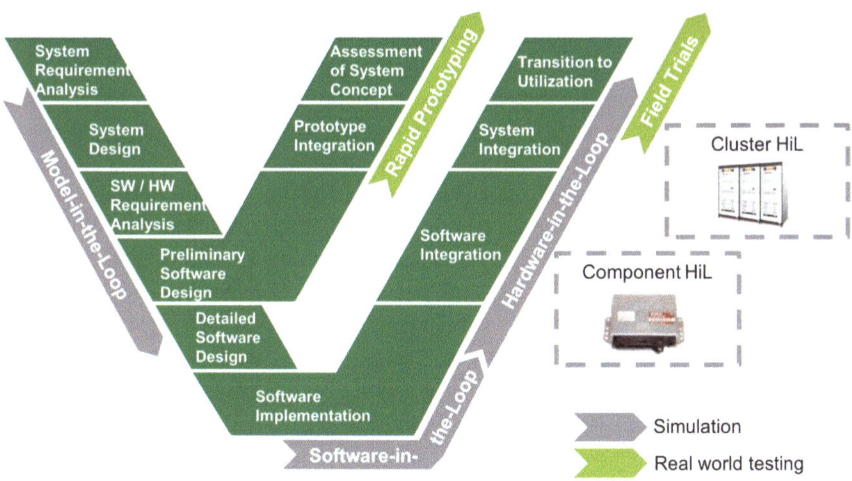

Fig. 1 V-Model with associated test technologies

for validation and release of automotive systems. The integrated mechatronic system is tested within the realistic environment under varying conditions, such as different weather or time of day. Scalability for the increasing number of scenarios as well as reproducibility are challenging especially regarding test of automated driving (Helmer et al. 2015).

X-in-the-loop (XiL) represents a widely used simulation approach in the automotive domain (Shokry and Hinchey 2009). The embedded software is tested against a plant model of the system environment within different development stages. Model-in-the-Loop (MiL) technology and prototyping are used in early development phases to validate the system concept (Shokry and Hinchey 2009). Commonly, MiL approaches are comprised within model based software development approaches for higher level of abstraction (Sax 2008). Software-in-the-Loop (SiL) tests are used to verify the implemented and integrated software modules within a virtualized runtime environment (Broekman and Notenboom 2002). On the right side of the V-Modell, the system is successively integrated. Starting with software components, which integrate into corresponding hardware in terms of an Electronic Control Unit (ECU), finally integrating into the entire vehicle. Hardware-in-the-loop (HiL) describes the functional test of the software integrated on the target hardware based on the usage of technical interfaces (Shokry and Hinchey 2009). A distinction is done between Component HiL and Cluster HiL. Within the Component HiL, a single ECU is tested regarding its correct functionality (Schäuffele and Zurawka 2012). Cluster HiL serves for verification of complex and interconnected vehicle systems. The aim of this test stage is to systematically cover the entire data flow and chain of effects regarding an entire feature. The chain of effect consists of several sensors, functional elements, and actuators (Wohlfahrt et al. 2016).

Figure 2 shows the schematic architecture of a Cluster HiL system that is used for testing of driver-assistance systems. The Cluster HiL consists of two parts. The HiL simulator that manages the operation of electronic control units, actuators, and sensors, like the stereo camera test bench. The second part contains the simulation environment. It comprises the simulation of the virtual world and image generation as well as a scenario-database for the driving scenarios.

Actual test cases for HiL systems are systematic tests focusing on requirements coverage (Wohlfahrt et al. 2016). Test cases are specified following a common structure: an initial state of the system (precondition), an order of actions and events within the test case (test steps) and expectations for the outcome (conditions) are expressed. Therefore, the test case comprises a single driving scenario. Within this scenario, predefined actions are triggered based on timing or events. The evaluation of the test cases is discrete in form of pass or fail. Therefore, the entire system behavior must be known a priori (Shokry and Hinchey 2009). Test catalogs are used to cover the entirety of requirements and contain different test cases (International Organization for Standardization 2017). For each requirement, at least one test case is defined to satisfy requirement coverage regulations.

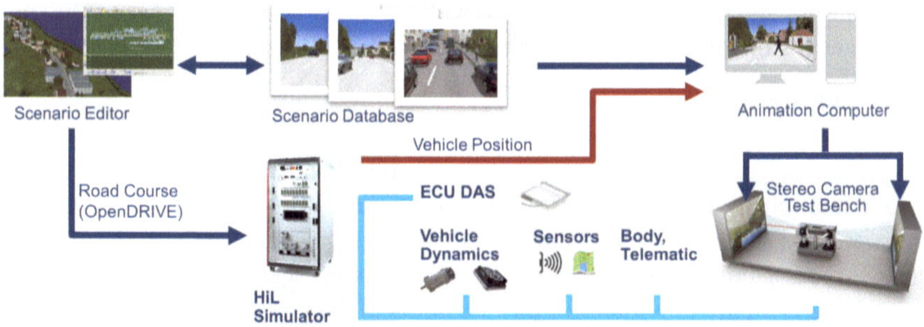

Fig. 2 Schematic Cluster HiL architecture for driver-assistance systems (Wohlfahrt 2016)

2.2 *Digital Test Drive*

Due to the continuous expansion of the variety of driving scenarios and relevant situations required for testing of automated driving systems, the necessary amount of test kilometer increases steadily (Wachenfeld and Winner 2016). Real-world test drives are no longer sufficient to cover the resulting test amount and efforts for simulation and virtual testing will increase (Wachenfeld and Winner 2016). Since virtual tests of assistance systems are mostly based on XiL technology there is a wide range of simulation approaches and tools. These approaches reach from macroscopic traffic simulation with a global perspective (Krajzewicz et al. 2012) to microscopic driving simulation (Von Neumann-Cosel et al. 2009) as well as augmented reality applications (Zofka et al. 2014) with egocentric perspectives. Digital test drives represent the next step of virtualization complementing systematic testing with XiL and aiming on taking realistic field trial test drives into simulation environments. The basic idea is to model a test drive with a parametrizable and realistic driver and traffic behavior model (Wohlfahrt et al. 2016). The digital test drive can be used to evaluate complex driving scenarios and situations entirely, comprising existing specification-based tests. The digital test drive is based on the infrastructure of the Cluster HiL for simple integration. As stated above, for entire driving scenarios novel methodologies for the assessment of the system-under-test are required. An initial concept for the automated assessment is given in this contribution.

3 Requirements and Constraints for Automated Assessment of Digital Test Drives

In order to achieve an automated assessment for a digital test drive, it is essential to identify necessary requirements and constraints. Therefore, current test approaches and stages regarding Cluster HiL and real-world tests have been analyzed regarding their specification and application as well as their aims and goals. The most important results and aspects are elaborated in the following:

Continuous evaluation
As already mentioned, it is important to enable a continuous assessment of the system-under-test. The dynamic behavior of the digital test drive is not known a priori. In contrast to systematical tests, a digital test drive does not work with test steps or fixed states. Therefore, approaches for the evaluation of the entire test drive are required. Additionally, the recognition of driving scenarios and maneuvers of the vehicle is highly important for the activation of appropriate assessments within a test drive.

Asynchronous and parallel assessment
To reduce expenditure of time, it is important to run multiple assessments in parallel without any restrictions and influences. For the same reason, it should be possible to evaluate multiple systems and functions in parallel.

Conjunction of different types of assessment
It is mandatory to facilitate statements about different test criteria like applicable laws or function-specific aspects. In this context, the abstract description of driving scenarios (Bach et al. 2016) gains importance. An abstract description of driving scenarios enables to reuse one scenario recognition for different assessments. This reduces the effort and complexity of the actual realization of the assessment concept.

Scalability
Our research reveals that the scalability of the concept is one of the most important requirements. The concept needs to be applicable for relatively simple functions such as lane departure warning but also for complex systems like advanced driver-assistance systems without further modifications of the concept.

4 Automated Assessment Concept

To cover the described requirements and constraints, we propose an automated assessment concept for digital test drives as shown in Fig. 3. The test concept supports evaluation of digital test drives comprising established systematic tests with XiL to ensure adequate level of maturity of the system-under-test.

The concept consists of four different domains: the HiL system, the automated assessment domain, the visualization domain, and the data analytics domain. To foster reusability and simple design of assessments, activation conditions are distinct from test conditions. The assessments are grouped into three main categories to allow different assessment focus. These include the general assessment as well as system-specific aspects and the assessment of the test system. Novel conditions and assessments can be derived based on exploratory data analysis. The extraction of relevant context information for the activation conditions such as the actual

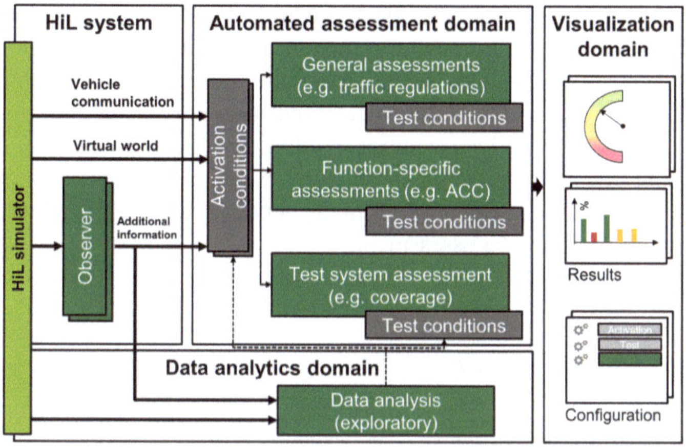

Fig. 3 Automated assessment concept for digital test drives

maneuver state of the vehicle is achieved by observers. During a digital test drive, not all assessments are active all the time. Some assessments are bound to specific functions, the occurrence of situations or vehicle maneuver states. The concept with its domains will be presented in detail in the following sections.

4.1 HiL System

The HiL system comprises two different functional modules. On the one hand, the HiL simulator including all its components as introduced in Chap. 2, which includes all vehicle communication as well as the simulated environment. On the other hand, there are additional observers. The concept of observers is related to control theory, where an observer aims at providing information on the internal state of the system based on measurements (Khalil 2002). For the purposes of this contribution, the observer is a software module that interprets continuous data from the virtual world with the aim to derive additional information. Exemplarily, an observer identifies the actual driving maneuver of the ego-vehicle or other surrounding vehicles within the test drive. This could lead to the identification of driving scenarios exemplarily based on openScenario[1] or related modeling approaches (Bach et al. 2016). Hence, data provided by the HiL system is interpreted to provide information gain for the automated quality assessment on different abstraction levels and with different focus.

The environment information encompasses information about the virtual world, like the road type, applicable laws or weather as well as information about

[1]http://www.openscenario.org.

surrounding road users. Generally, there are two major types of information: raw information and derived additional information. The term *"raw information"* refers to non-interpreted data from the vehicle communication bus and the simulation of the virtual world. The utilized full vehicle simulation provides all information by a runtime data bus. The derived additional information is the output of integrated observers and comprises processed data such as the maneuver states of the participants. Moreover, the information from the simulation of the virtual world can be used as ground truth for the assessment.

4.2 Assessment Domain

In the automated assessment domain, the actual assessment is performed. This contribution proposes an approach that divides every assessment in two parts: the activation of the assessment and the actual assessment itself. Both the activation and assessment are composed of conditions, namely activation and test conditions. Furthermore, it is important to emphasize that this approach includes an abstraction concept for activation conditions. The goal behind using abstracted conditions is to simplify the handling and increase the reusability of the activation condition for different assessments.

We propose a distinction of the assessment into three categories. The first category comprises general and overall assessments, such as applicable legal aspects regarding traffic regulations, assessment programs like European New Car Assessment Program (Euro NCAP) or basic laws of physics. The second category includes function-specific verifications. They are used to evaluate the functionality of driver assisting systems like Adaptive Cruise Control (ACC), Lane Keeping Assistant (LKA) or PreSafe Levels (Bogenrieder et al. 2009). The last assessment category focuses on the test system. Assessments in this category evaluate the relevance of the statistically generated digital test drive. Moreover, coverage of different experienced driving scenarios and situations within the digital test drive are assessed for statistical relevance and coverage.

4.3 Visualization and Data Analytics Domain

Within the visualization domain, the assessment results are prepared for visualization. A graphical user interface supports displaying the results and configuring the assessments. The visualization comprises the assessment results as well as the distribution of test scenarios.

The fourth domain focuses on data analytics. As the digital test drives produces massive unknown test data, novel methods for the analysis are required in order to gain knowledge about undiscovered scenarios of high relevance. This domain is a research area whose potential is currently not fully evident. Exploratory data analysis methods are under investigation for application. This will enable to detect new activation conditions or anomalies within the digital test drive.

5 Application on Exemplary Driver-Assistance System

To demonstrate the application of our concept, Fig. 4 shows a basic assessment for an ACC system. Derived from the requirements of the ACC, the system should perform either a velocity control or a time gap control, depending on the actual maneuver state of the vehicle. We consider the maneuver *"free cruise"* and *"follow"* in this example, the maneuver *"approach"* is not elaborated but can be included as an additional assessment without large expense. The maneuver state is interpreted by the dedicated observer, based on the simulation environment data. The data flow is given from left to right and the necessary information for the activation of the assessments are outlined in the components. The HiL system also provides information for the test conditions based on the vehicle network signals in terms of the activation state of the different assistance systems.

The example comprises two different function-specific assessment of the ACC. These can be triggered in parallel, based on the fulfillment of different activation conditions. Additionally, general assessments such as space of freedom and lane keeping are active during the entire test drive, but are not in focus within this contribution. The first assessment aims at the evaluation of the *"free cruise"* maneuver of the ACC and the second assessment targets the *"follow"* maneuver. To activate the first assessment, the maneuver state of the vehicle needs to be *"free cruise"* and the ACC needs to be activated. If both statements are fulfilled, the activation condition is true and the tests conditions are valid. For assessment one, we modeled two test conditions. The first test conditions checks whether the ACC status is correctly displayed in the instrument cluster. The second test checks the actual velocity control and if the target velocity is reached after a certain time. Due to the nature of the activation condition, the

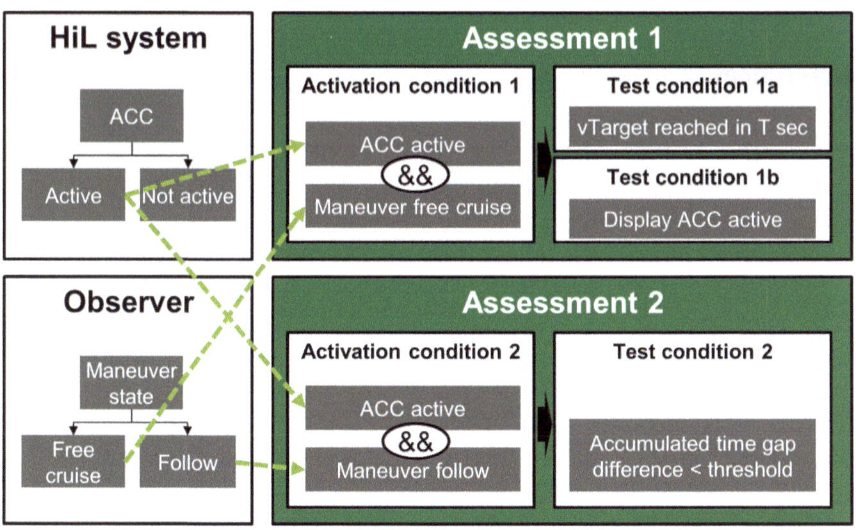

Fig. 4 Example application for the assessment of an adaptive cruise control

assessment can be triggered multiple times during a digital test drive. This refers to a driving scenario being included several times within the test drive.

Additionally, an assessment for the "*follow*" maneuver is performed. The activation condition is similar to the first one but this time the maneuver state of the vehicle needs to be "*follow*". The corresponding test conditions checks whether the controlled time gap is within a designated range around the desired value. Both assessments are active in parallel and allow multiple assessments during one scenario of the digital test drive. The proposed assessment concept enables a systematic evaluation of statistically generated test scenarios. The overall verification and validation activities benefit from a wider coverage due to the evaluation of several situations during one test scenario of the digital test drive.

6 Conclusion and Outlook

In this contribution we presented an approach for the automated assessment of digital test drives. The concept enables the automated assessment without predefined test scenarios as well as parallel assessments of different systems and functions. The assessment of digital test drives complements systematic tests on the Cluster HiL infrastructure and contributes significantly to the extending effort for virtual verification and validation. As this contribution provides an initial prove of concept, we will fully implement and integrate the concept to prove scalability within the next steps. Methodical research will address the transition between different maneuvers as well as possibilities for exploratory data analysis to derive undiscovered scenarios and assessments.

References

Bach J, Otten S, Sax E (2016) A model-based scenario specification method to support development and test of automated driving functions. In: IEEE Intelligent vehicles symposium (IV)

Bach J, Otten S, Sax E (2017) A taxonomy and systematic approach for automotive system architectures: From functional chains to functional network. In: 3rd International conference on vehicle technology and intelligent transport systems

Bengler K et al (2014) Three decades of driver assistance systems. IEEE Intell Transp Syst Mag 6 (4):6–22

Bogenrieder R, Fehring M, Bachmann R (2009) Pre-safe in rear-end collision situations. In: 21st International technical conference on enhanced safety of vehicles

Broekman B, Notenboom E (2002) Testing embedded software. Addison-Wesley

Helmer T et al. (2015) Safety performance assessment of assisted and automated driving by virtual experiments: Stochastic microscopic traffic simulation as knowledge synthesis. In: IEEE International conference on intelligent transportation systems (ITSC)

International Organization for Standardization (2017) ISO 29119, Software and systems engineering—Software testing—Part 2: Test processes

Khalil HK (2002) Nonlinear systems. Prentice Hall, Upper Saddle River NJ 07458
Krajzewicz D et al. (2012) Recent development and applications of sumo simulation of urban mobility. Int J Adv Syst Meas
Sax E (2008) Automatisiertes Testen eingebetteter Systeme in der Automobilindustrie. münchen. Carl Hanser Verlag, München Wien
Schäuffele J, Zurawka T (2012) Automotive software engineering—grundlagen, prozesse, methoden und werkzeuge effizient einsetzen. Springer, Wiesbaden
Shokry H, Hinchey M (2009) Model-based verification of embedded software. Computer 42 (4):53–59
VDA QMC Working Group 13 and Automotive SIG (2015) Automotive SPICE Process Assessment/ Reference Model, 3rd edn. Berlin, Germany. http://www.automotivespice.com
Wachenfeld W, Winner H (2016) The release of autonomous vehicles. In: Autonomous Driving. Springer, pp 425–449
Weber J (2009) Automotive development process. Springer-Verlag
Wohlfahrt C et al. (2016) Von Systematischer Absicherung zur ‚Digitalen Erprobungsfahrt, 6. Fachkonferenz AUTOTEST Stuttgart
Von Neumann-Cosel K, Dupuis M, Weiss C (2009) Virtual test drive–provision of a consistent toolset for [d, h, s, v]-in-the-loop. In: Driving simulation conference Europe
Zofka M et al. (2014) Semivirtual simulations for the evaluation of vision-based adas. Intelligent Vehicles Symposium

HiFi Visual Target—Methods for Measuring Optical and Geometrical Characteristics of Soft Car Targets for ADAS and AD

Stefan Nord, Mikael Lindgren and Jörgen Spetz

Abstract Advanced Driver-Assistance Systems (ADAS) and Automated Driving (AD) vehicles rely on a variety of sensors and among them optical sensors. Extensive testing of functions using optical sensors is required and typically performed at proving grounds like AstaZero. Soft surrogate targets are used for safety reasons but the optical and geometrical characteristics of soft car targets may differ considerably from that of real vehicles. During tests the quality of the soft car targets deteriorates due to repeated impacts and reassembly of the targets, and there is a need of methods for securing the quality of the soft car targets over time. One of the main goals of the HiFi Visual Target project is to develop and validate accurate and repeatable measurement methods of the optical and geometric characteristics of soft car targets.

Keywords Advanced driver-assistance systems · Automated driving · Active safety · EuroNCAP · Vision zero · Soft car target · Optical sensors · Measurement technology · Spectroradiometer · Spectral reflectance · Geometrical scan

1 Background

There are three grand challenges the transportations of tomorrow face: environment, safety, and congestion. One key element in meeting these challenges, and also to reach the VisionZero stated by the Swedish government in 1997, is the development of active safety systems and AD (Automated Driving) systems assisting or replacing the driver in both normal traffic situations as well as critical situations.

S. Nord (✉) · M. Lindgren · J. Spetz
Safety and Transport, Measurement Science and Technology, RISE Research Institutes of Sweden, Box 857, 50515 Borås, Sweden
e-mail: stefan.nord@ri.se

M. Lindgren
e-mail: mikael.lindgren@ri.se

J. Spetz
e-mail: jorgen.spetz@ri.se

C. Zachäus et al. (eds.), *Advanced Microsystems for Automotive Applications 2017*, Lecture Notes in Mobility, DOI 10.1007/978-3-319-66972-4_17

These systems have already proven to decrease the number and severity of injuries and insurance cost (Volvo car corporation leads the way in car safety: risk of being injured in a Volvo reduced by 50 % since the year 2000; Isaksson-Hellman and Lindman 2015). The development of higher levels of AD puts stronger requirements on reliability, and therefore on its sensory input and the interpretation of this input. There exist several sensor types that typically are employed in sensor systems in ADAS and AD, such as visible spectrum and infrared cameras, laser scanners, ultrasonic sensors, and radars. In the HiFi Visual Target project the focus is on the optical sensors. Validation of the optical sensor systems and identification of possible performance issues require extensive testing of the systems, both during optical sensor system development, and during verification of ADAS and AD vehicles. Controllability and safety of testing dictates that most testing is typically performed on closed test tracks.

2 Soft Car Targets

Since it is not possible to test situations that may result in collision with real vehicles, pedestrians, etc., as targets, surrogate objects are used, see Fig. 1. These are typically mock-up objects made of soft materials that can be repeatedly hit without damage to themselves or the test vehicle (Rabben et al. 2015). In order to make testing with surrogate targets to be valid, sensor response of the surrogate target must be consistent with the response of the corresponding real target. While several studies have been conducted where sensor responses of cars and humans have been measured (Sandner 2013; Yamada 2001; Suzuki 2000; Yamada 2004), most of the work has either been concentrated on radar sensor responses (RCS properties) of cars and humans or the optical characteristics of human targets (Lemment et al. 2013). There is lacking a more thorough work on the optical and geometrical characteristics of soft car targets. Most of the work has also been

Fig. 1 Soft Car Targets (SCT) are used for safe and repeatable evaluation of ADAS and AD

focused on the rear-end aspects of car targets, but analysis of rear-end aspects only provides validation for rear-end scenarios. The aspects that might emerge in more complex traffic scenarios, for example, intersection collision avoidance (City safety by Volvo Cars—outstanding crash prevention that is standard in the all-new XC90 2014), or merging, overtaking and cut-in cases for AD vehicles, where the optical sensor responses on target vehicles may vary rapidly, are not covered, limiting the applicability of the studied surrogate target for testing of these situations.

Moreover, analyses of other types of surrogate targets available on the market, as well as studies of possible improvements of the latter, are lacking. This poses limitations for OEMs, optical sensor developers and test tracks who need reliable surrogate targets for increasing need for reliable and safe optical sensor testing.

There is a need to accelerate development and minimize conflicting requirements for test targets that vehicle manufacturers, suppliers, legislators, and consumer organizations can use for the safe and repeatable evaluation of vehicles. Euro NCAP (European New Car Assessment Programme), NHTSA (National Highway Traffic Safety Administration), and IIHS (Insurance Institute for Highway Safety) are collaborating on specifications and assessing designs for a harmonized future 3D vehicle test target that will be compatible with a variety of target carriers. Additionally, there is a need of methods for validation that test targets fulfills its specifications over time and that any method developed for use on the test track must be fast and easy to use in order to save time and money as equipment and test track are big investments that call for efficient use for high productivity.

3 Project Goals

The objective of the project is to enable more efficient and reliable verification of optical sensor systems, including ADAS and AD systems that rely on the optical sensors, through:

- development and validation of accurate and repeatable measurement methods of the optical and geometrical characteristics of soft car targets,
- providing input to the development of more realistic soft car targets for safe testing of automotive optical sensor systems,
- demonstration of improved verification with the developed measurement methods,
- supporting international standardization (ISO) with standard methods enabling future verification and calibration of optical characteristics of soft car targets.

The ambition in the project is to contribute to these goals through delivering (i) test methods and equipment specifications for securing the optical and geometric properties of soft car targets over time (ii) proposal on improvements of high fidelity soft car targets that significantly enhance current state of the art. Throughout the project, we will:

- investigate what test setup is needed for measurement of optical and geometric characteristics of automotive targets with a focus on cars,
- analyze the optical and geometric characteristics of real targets and introduce requirements for the characteristics that soft car targets shall have. We will assess the soft car targets and evaluate their resemblance with real counterparts for optical sensors, and
- propose improvement of soft car targets to match requirements for target characteristics.

4 Initial Measurements and Results

4.1 Measurement Setup

Two SCTs (Soft Car Targets) of the same type were studied, one in near mint condition and one heavily used, see Fig. 2. Two different setups were used to study different properties of the SCTs; one optical setup and one geometry-measuring setup.

4.1.1 Optical Measurement Setup

In the optical setup, the surface reflectance of the SCTs was measured by means of a spectroradiometer and a typical Autoliv vehicle camera system. The SCTs were placed in an enclosed space with covered windows. The reflectance was measured in ~60 different points all around the SCT. In Fig. 3, each measurement point is marked with a small black pointer.

An incandescent light source was used to illuminate the SCT for measurement. The light source was placed 10 m from the SCT and at an angle of ~30° relative to the horizontal plane. The SCT was rotated around its vertical axis into 12 different positions (30° interval) for measurement to take place at roughly normal incidence to the selected measurement spot. The spectroradiometer was placed 5 m from the target and used a 1° measurement field-of-view, equivalent to a spot size of ~88 mm on the SCT surface.

Fig. 2 The used SCTs, *left* is in near mint condition and *right* has been heavily used

Fig. 3 Example of the image collected by the vehicle camera system. Four *black pointers* on the SCT side indicate the position of measurement spots for the spectroradiometer. Also visible on the floor are the markings for *center* position and rotation of the SCT

4.1.2 Geometry Measurement Setup

The geometry was measured by means of a laser scanner, capable of generating a point cloud with three-dimensional information with an accuracy down to millimeters. Figure 4 shows the setup with the laser scanner and spectroradiometer.

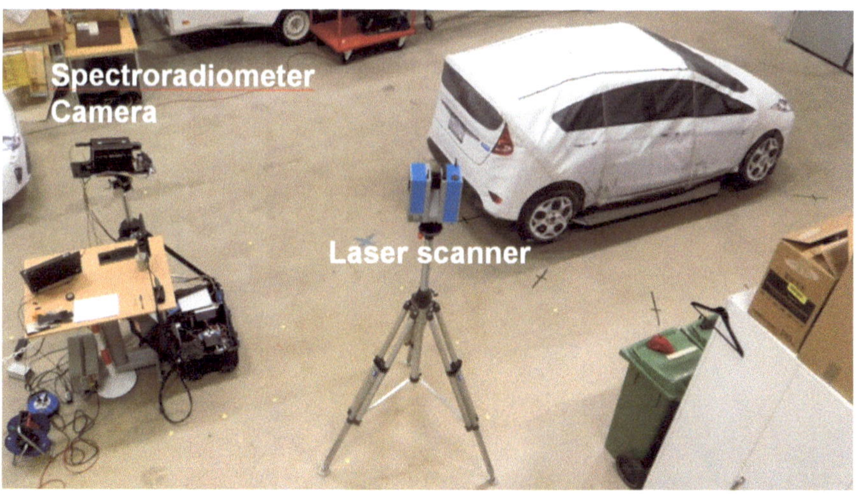

Fig. 4 Measurement setup with the laser scanner and the spectroradiometer. The SCT is in the position used for geometry measurement

4.2 Preliminary Results

At the time of writing, the results of measurement are still being evaluated and preliminary results are presented here.

4.2.1 Optical Measurement Results

The reflectance for the measured spots is shown in Fig. 5.

The data shows a considerable resemblance between the measurements made on two different SCTs, one heavily used and one in near mint condition, indicating that the effect of hitting the SCT does not degrade the material very quickly. Also in the infrared wavelength region, which is used by laser radar (typically 905 nm), there is not a significant change in appearance.

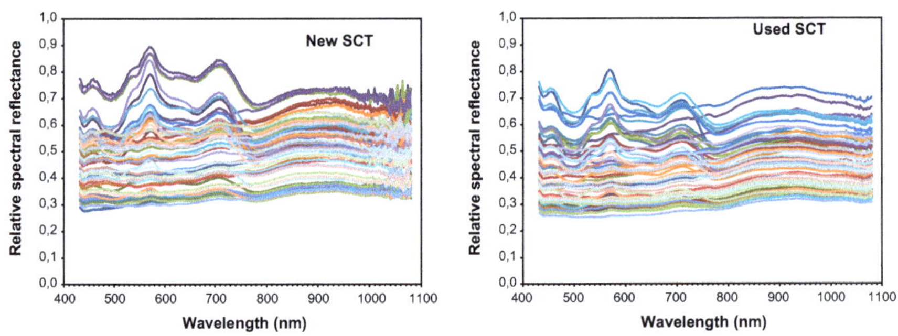

Fig. 5 Raw data for spectral reflectance. Further analysis is required to determine the relevant differences between the new and used SCT

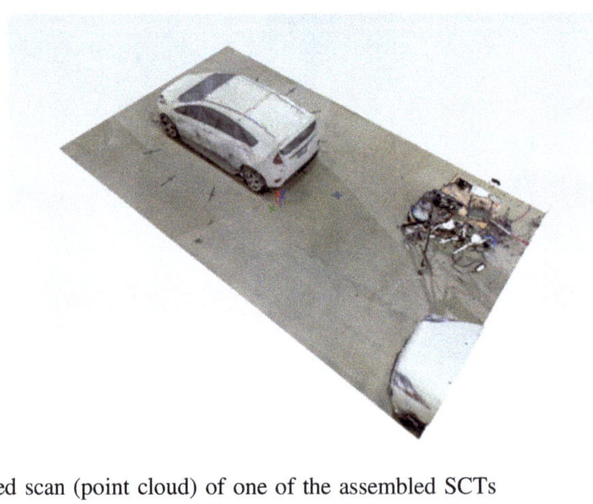

Fig. 6 Registered scan (point cloud) of one of the assembled SCTs

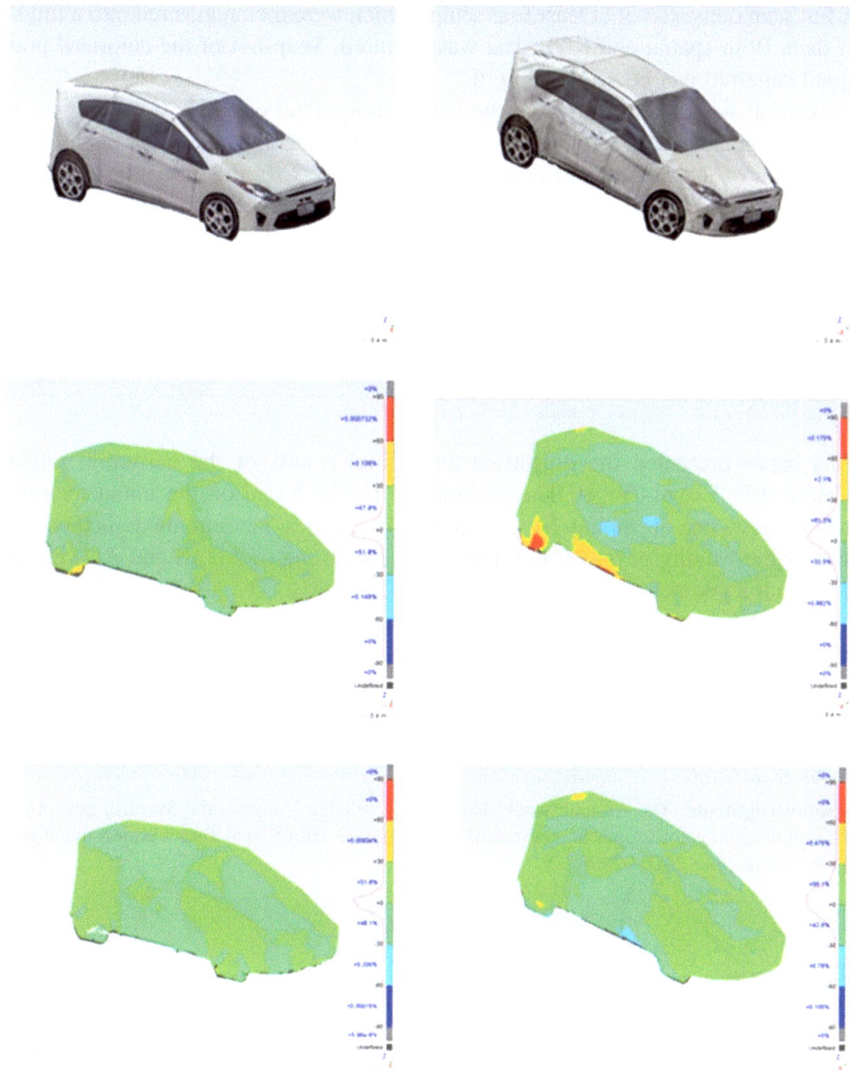

Fig. 7 *Left* column shows result for the new SCT, *right* column for the used SCT

4.2.2 Geometry Variation Due to Assembly

The SCTs consist of a structure that requires careful assembly. Due to their design and the fact that they can be used on different target carriers, variations in geometry are obviously to be expected. In order to understand what impact the assembly will make on the geometry, both SCTs were assembled and disassembled several times. For each step, they were scanned in full using a TLS (Terrestrial Laser Scanner).

A full scan consisted of at least four setups which were then registered into a full set of data. Both spatial and RGB data was captured. Snapshot of the colorized point cloud captured can be seen in Fig. 6.

A mesh was constructed from the initial point cloud of each SCT acting as the reference geometry. The following datasets were then compared to the initial mesh. The result of these comparisons can be seen in Fig. 7. In this initial test, different carriers (as used by the respective owner) were used. Therefore, the comparison covers only the body of the SCT itself. The initial analysis shows, as expected, larger variations for the heavily used SCT.

5 Conclusions and Future Work

This paper presented the objectives and initial results of the currently running project HiFi Visual Target. The preliminary results based on the initial measurements show that it should be possible to develop a measurement method that secures the validity of Soft Car Targets used for testing of ADAS and AD vehicles from both a geometrical and optical sensor perspective. The geometry of the target could, e.g., be specified in a way that could simplify the verification of the shape in order to secure that it fulfills its specification, even after long time usage and a large number of assembly and disassembly. Future work includes further analysis of measurement data, measurements on real vehicles and wearing out an SCT while applying developed measurement methods at the AstaZero proving ground.

Acknowledgments The Authors would like to acknowledge Vinnova, the Swedish government innovation agency, for supporting these activities within the HiFi Visual Target project under grant agreement number 2016-02496.

References

City safety by Volvo Cars—outstanding crash prevention that is standard in the all-new XC90, press release from 2014, December 5. https://www.media.volvocars.com/global/en-gb/media/pressreleases/154717/city-safety-by-volvo-cars-outstanding-crash-prevention-that-is-standard-in-the-all-new-xc90. Accessed 18 May 2017

Isaksson-Hellman I, Lindman M (2015) Real-World performance of city safety based on swedish insurance data. In: 24th international technical conference on the enhanced safety of vehicles (ESV), no. 15-0121, Gothenburg, Sweden, 8–11 June 2015

Lemment P et al (2013) Evaluation of pedestrian targets used in AEB testing: a report from harmonisation platform 2 dealing with test equipment. In: 23rd international technical conference on the enhanced safety of vehicles (ESV), no. 13-0124, Seoul, Korea, South, 27–30 May 2013

Rabben M, Henze R, Küçükay F (2015) Dynamic crash target for the assessment, evaluation and validation of ADAS and safety functions. In: Proceedings of the 3rd international symposium on future active safety technology towards zero traffic accidents. (and references therein)

Sandner V (2013) Development of a test target for AEB systems. In: 23rd international technical conference on the enhanced safety of vehicles (ESV), no. 13-0406, Seoul, Korea, South, 27–30 May 2013

Suzuki H (2000) Radar cross section of automobiles for millimeter wave band. JARI Res J 22 (10):475–478

Volvo car corporation leads the way in car safety: risk of being injured in a Volvo reduced by 50 percent since year 2000. Press release. https://www.media.volvocars.com/global/en-gb/media/pressreleases/45468. Accessed 18 May 2017

Yamada N (2001) Three-dimensional high resolution measurement of radar cross-section for car in 76 GHz band. R&D Rev Toyota CRDL 36(2). (Toyota Central R&D Labs, Inc)

Yamada N (2004) Radar cross section for pedestrian in 76 GHz. In: Proceedings of 2005 European microwave conference, vol 2, pp 46–51, 4–6 Oct 2004

Part V
Legal Framework and Impact

Assessing the Impact of Connected and Automated Vehicles. A Freeway Scenario

Michail Makridis, Konstantinos Mattas, Biagio Ciuffo, María Alonso Raposo and Christian Thiel

Abstract In the next decades, road transport will undergo a deep transformation with the advent of connected and automated vehicles (CAVs), which promise to drastically change the way we commute. CAVs hold significant potential to positively affect traffic flows, air pollution, energy use, productivity, comfort, and mobility. On the other hand, there is an increasing number of sources and reports highlighting potential problems that may arise with CAVs, such as, conservative driving (relaxed thresholds), problematic interaction with human-driven vehicles (inability to take decisions based on eye contact or body language) and increased traffic demand. Therefore, it is of high importance to assess vehicle automated functionalities in a case-study simulation. The scope of this paper is to present some preliminary results regarding the impact assessment of cooperative adaptive cruise control (CACC) on the case-study of the ring road of Antwerp, which is responsible for almost 50% of the traffic and pollution of the city. Scenarios with various penetration rates and traffic demands were simulated showing that coordination of vehicles may be needed to significantly reduce traffic congestion and energy use.

Keywords Cooperative adaptive cruise control (CACC) · Connected and automated vehicles (CAVs) · Platooning · Traffic simulation · Traffic flow

---

M. Makridis (✉) · K. Mattas · B. Ciuffo · M.A. Raposo · C. Thiel
Directorate for Energy, Transport and Climate Change, European Commission – Joint Research Centre, Via E. Fermi 2749, 21027 Ispra, VA, Italy
e-mail: michail.makridis@ec.europa.eu

K. Mattas
e-mail: konstantinos.mattas@ec.europa.eu

B. Ciuffo
e-mail: biagio.ciuffo@ec.europa.eu

M.A. Raposo
e-mail: maria.alonso-raposo@ec.europa.eu

C. Thiel
e-mail: christian.thiel@ec.europa.eu

C. Zachäus et al. (eds.), *Advanced Microsystems for Automotive Applications 2017*,
Lecture Notes in Mobility, DOI 10.1007/978-3-319-66972-4_18

1 Introduction

In the next decades, road transport will undergo a deep transformation with the advent of connected and automated vehicles (CAVs), which are about to drastically change the way we commute. CAVs hold significant potential in the reduction of road accidents, traffic congestion, traffic pollution, and energy use (Litman 2015), promising to increase productivity and comfort and to facilitate a greater inclusion in mobility of specific user groups, which may eventually lead to increased travel demand. Together, CAVs will enable the full potential of self-driving technology and they will completely merge over time in the next 30 years (Alonso Raposo et al. 2017). The complexity of transportation systems is high, and therefore efficient tools for the assessment of this disruptive change are important.

The cooperative adaptive cruise control (CACC) is a completely automated driving functionality, similar to the driving behavior that we expect autonomous vehicles to have at least on the level of V2V communication. Furthermore, since transportation phenomena are complex and highly not linear, the effect of CAVs on a complete road network cannot be evaluated exhaustively with simulations on single sections, links, or merging points. It is important to assess these automated functionalities in a case-study using as realistic as possible traffic dynamics and as similar as possible vehicle movement logic. Hence, the scope of this paper is a preliminary assessment of CACC technology on a case-study scenario of a large network based on real data and performing simulation tests for variable CAVs penetration rates, and variable traffic demands. Furthermore, using detailed results of the simulation tests, it is possible to measure not only network performance characteristics, but also energy consumption and environmental impacts. Case study for the present work is the ring road of Antwerp (a city with a population over half a million) which is responsible for almost 50% of the traffic and pollution of the city (Lefebvre et al. 2011; Lefebvre et al. 2013; Degraeuwe et al. 2016). The simulations are carried out using Aimsun, a commercial simulation software, customizing the various scenarios using its API capabilities.

The rest of the paper is organized as follows: Sect. 2 presents the relevant literature regarding simulation of CACC components; Sect. 3 describes the network, data and algorithms used for the simulation scenarios; Sect. 4 discusses the results of the simulation and finally conclusions and future work are presented in Sect. 5.

2 Review of the Literature

In the literature, CACC has been tested using microscopic simulation tools to evaluate its impact on traffic flow, road capacity, and travel time. The scenarios that have been tested so far include different market penetrations of vehicles having CACC capability, as well as different schemes to model the behavior of CACC on single-lanes or multi-lane sections, on merging sections, or in intersections (Ilgin

Guler et al. 2014). Experimental results state that even at low penetration rates of CACC, the capacity of a section is increased. Those resulting new capacities have been adopted by researchers, and embedded on macroscopic simulations, to explore the new network characteristics (Ngoduy 2013; Fakharian Qom et al. 2016; Kloostra and Roorda 2017). However, to the best of the authors' knowledge, no microsimulation approach has presented results on anticipated increase in the traffic demand, although any change will immediately reflect on the overall performance of the road network.

Furthermore, in the literature, for the microscopic simulation of CAVs in 2006 van Arem et al. (2006) proposed a CACC model on one section of a Dutch highway with a lane drop from 4 to 3 lanes, finding that CACC can improve the capacity and soften the observed shockwaves on highways. They also concluded that a CACC dedicated line becomes beneficial once CACC penetration is higher than 40%. In 2011, Shladover et al. (2012) created a mixed scenario with manual vehicles, ACC, and CACC-equipped vehicles to test the change on lane capacity. They also changed the desired time gap, according to data from field tests, concluding that as long as the driver chooses the time gap, there is a trend to choose comfortable time gaps, so the network can benefit enough from neither CACC nor ACC, as their capabilities are restrained. Calvert et al. (2012) simulated two connected freeways in Amsterdam. Arnaout and Bowling (2014) simulated a highway section with an on ramp for multiple CACC market penetration rates. Chin et al. (2015) advocated the inaccuracy of many models to simulate human driver behavior and the possibly misleading effects of testing mixed human and CAVs simulation with such models. Therefore, they proposed a new model (PrARX) that predicts human driver behavior probabilistically and adjusts CACC behavior to those predictions. Xiao et al. (2017) summed up existing CACC simulation models and proposed a new one, adopting a human overtake at hazardous situations. In 2016, Lu et al. (2017) also proposed their own driver model to unify human-driven and C/ACC vehicles, calibrated by data from freeway in Sacramento, California. To sum up, the case studies presented in the literature, mainly focus on macroscopic traffic simulation or the microscopic traffic simulation of relatively small networks. In addition, they advocate the *ceteris-paribus* principle, assuming that everything else but the introduction of CAVs stays constant (e.g., the traffic demand). In this light, the objective of the present paper is to study the effect of different penetration levels for CAVs and of different scenarios of demand evolution inside a complex case-study.

3 Case-Study Simulation

This section describes (a) the work carried out to develop the traffic model of the ring road of Antwerp, implemented in Aimsun, (b) the characteristics of the car-following behavior, headway, and CACC logic used, (c) the metrics used to assess the simulation results, and finally (d) the scenarios tested with variable penetration rates and traffic demands.

3.1 The Traffic Model of Antwerp's Ring Road

The ring road of the city of Antwerp, in Belgium, has been chosen as case-study for the analysis due to its peculiarities. The ring road traffic is composed by commuters entering exciting and crossing the city, and by heavy-duty vehicles connecting the second biggest European port with the rest of the continent. As a result, the ring road presents from congested to saturated traffic conditions for many hours during the day and is responsible for more than half of the overall pollutant emissions generated by road transport in the city. It therefore represents a suitable case-study to assess the potential benefits connected to the introduction of CAVs. The traffic simulation model of the ring road has been developed in Aimsun (Fig. 2). The network topology and some of the attributes have been automatically extracted from OSM (2017) and then further post-processed to correctly reproduce the geometry of sections, junctions, and lanes. For example, Fig. 1 presents how the network was processed in order to visually reflect what appears in Google maps (http://maps.google.com). Fine-tuning the network geometry has resulted very important, since many of the traffic bottlenecks occurring during the day often

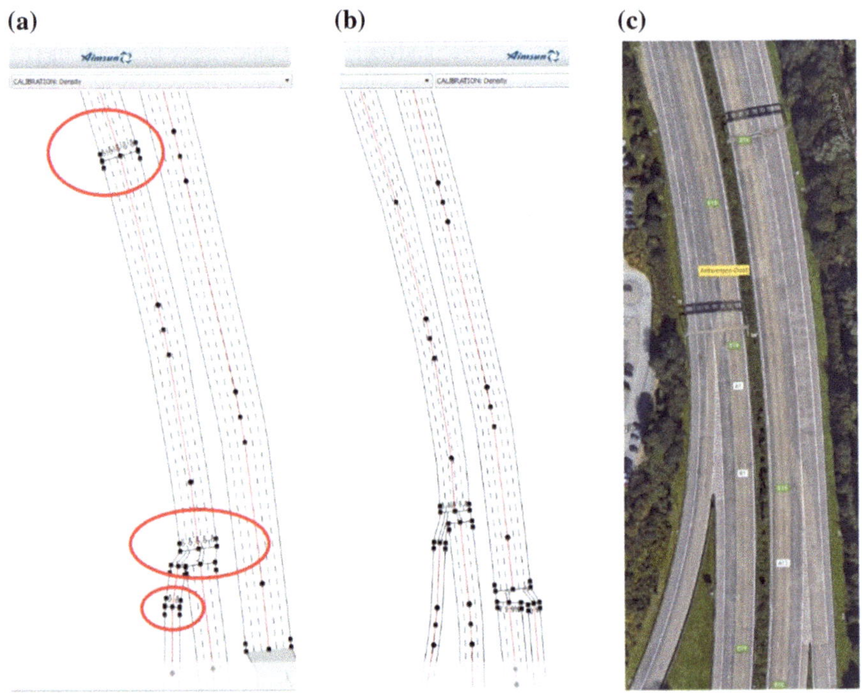

Fig. 1 Network fine-tuning **a** an example of some sections as as they were automatically loaded from OSM, **b** the same sections after processing in order to reflect what can be seen in google maps (**c**)

originate from the peculiarity of the road network. The final supply model of the network consists of 119 km of roads with 27 centroids (origin/destination points), 208 sections, and 117 intersections.

The traffic demand has been estimated by correcting an original O/D matrix in order to reproduce real traffic counts measured during the morning peak hours (Lefebvre et al. 2011; Lefebvre et al. 2013; Degraeuwe et al. 2016). The correction of the O/D matrix has been performed using a tool embedded in Aimsun and based on the Frankie and Wolfe algorithm (Frank and Wolfe 1956). In addition, in order to have a more realistic representation of the traffic demand, the simulation period has been extended so to have 1 h of network loading and 1 h of network discharging. In this way, it was possible to achieve a more natural traffic evolution.

3.2 *Human and CACC Drivers*

The Gipps model Gipps (1981) is used, as the default model to simulate human driver behavior, since it is widely accepted in the literature and it has been used in various papers.

The API of Aimsun was used in order to implement the CACC behavior proposed by Mahmassani (2016). The only simplifications made are in the sensor range, which is not limited in our simulation, and in the fact that every vehicle has the same maximum comfortable deceleration.

The acceleration ac_{cacc} is computed as follows:

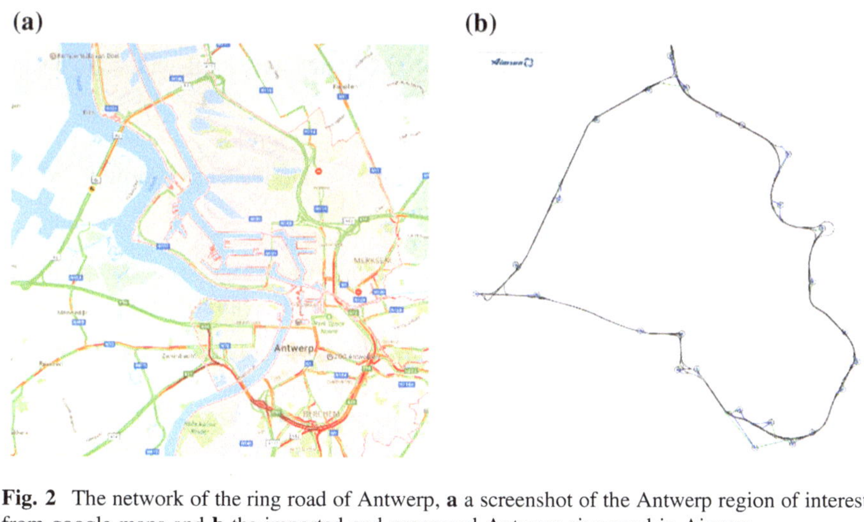

Fig. 2 The network of the ring road of Antwerp, **a** a screenshot of the Antwerp region of interest from google maps and **b** the imported and processed Antwerp ring road in Aimsun

$$ac_{cacc} = k_0 * ac_l + k_1 * (u_l + u) + k_2 * (g_o - g_d) \quad (1)$$

where k_0, k_1, k_2 constants decided to be 1, 0.5, 0.1 according to the literature, u_l and ac_l are the leader vehicle speed (m/s) and acceleration (m/s^2), g_o is the gap (m) that exists between the vehicles and g_d the desired gap of the follower vehicle (m).

Also acceleration must be inside the boarders of comfort as shown in the next equation:

$$ac = \max(-3, \min(2, ac_{cacc}, k * (u_l - u))) \quad (2)$$

Where constant k is equal to 1, as proposed in the literature, minimum, and maximum comfortable acceleration are −3 and 2 m/s^2. The actual acceleration is ac, ac_{cacc} is the acceleration calculated by Eq. (1), u_l is the leader vehicle speed (m/s) and u is the ego vehicle speed (m/s).

The maximum speed of the following vehicle is calculated to be safe if the leader decides to stop, using the maximum available deceleration. Also the minimum safe gap is calculated. Then the appropriate acceleration is computed, as follows:

$$u_{max} = \sqrt{u_l^2 - (g_o + u * SimStep) * 2 * \mathrm{b}} \quad (3)$$

$$g_d = \min(r_s * u, 2m) \quad (4)$$

Where u_{max} is the maximum speed (m/s), u_l is the leader vehicle speed and u is the follower vehicle speed (m/s)), g_o is the gap between the vehicles (m), $SimStep$ is the simulation step (sec), and b is the maximum deceleration of the ego vehicle (m/s^2), g_d is the minimum safe gap (m) and r_s a constant with value 0.5 according to the literature (sec). The derived speed cannot be more than the speed limit of the section, or the desired speed of the vehicle.

Finally for the CACC simulation, a set of constrains were implemented to decide where and when a vehicle can create a connection to its leader:

- Distance between the vehicles should be less than 200 m.
- Both vehicles must have CACC capabilities so the leader can provide to the follower the information.
- No information is sent from the follower to the leader.
- The two vehicles candidates for connection must be at the same lane and that their speed difference must not be more than a limit, arbitrarily set to 30 km/h.

If any of those constrains are not satisfied the vehicle is considered to be not connected and the behavior is ruled by the default Gipps model.

An important part of the driver model is the reaction time of the simulated vehicles. The human reaction time is quite long and automated functionalities tend to focus on reducing it, in order to increase capacity and reduce congestions. In the simulations the following reaction times were used:

Passenger cars: 0.8 s reaction time, 1.2 s reaction time at stop

Trucks: 0.8 s reaction time, 1.3 s reaction time at stop

Connected vehicles: 0.3 s reaction time addressing the safety gap, and the duration of mechanical processes.

It should be noted that the implemented simulation of CACC does not intervene on the Aimsun's default lateral movement behavior.

3.3 Assessment Metrics

For the assessment of the network's performance per simulation run, we use the harmonic average speed of the vehicles (m/s), the standard deviation of the speed (m/s), the average density of the network (veh/km), the average flow (veh/h), total energy consumption on the network based on the definition below (KWs). For each simulation scenario (involving variable traffic demand and penetration rate for CACC), the above values are used for the comparative analysis presented below in the results' section.

Energy Consumption

The method provided by Aimsun software in order to compute the fuel consumption of the vehicles in the network (European Environment Agency 2017) is considered quite old. Furthermore, fuel consumption and emissions are highly dependent on the ratios of diesel, fuel, electrical engines, etc., that the vehicle fleet uses, which can only be projected for future scenarios, adding unwanted uncertainty to the experiment. Hence, a different approach was used here and it was decided to calculate the energy consumption at the wheels from the vehicles flowing on the network. This metric is more robust as it is independent of the vehicle powertrain.

The formulation used in the present study to calculate the energy required by the vehicle is presented in the Eq. 5 below:

$$E = \sum_{t=o}^{T} P_t \Delta t = \sum_{t=o}^{T} \left(F_0 + F_1 v_t + F_2 v_t^2 + 1.03 m a_t\right) v_t \Delta t \tag{5}$$

where F_0 (N), F_1 (N*s/m), and F_2 (N*(s/m)^2) are the Road Load Coefficients, which describe the relationship between overall resistances to motion and the vehicle speed, m is the vehicle mass (kg), v_t and a_t are the speed (m/s) and acceleration (m/s^2) of the vehicle at time t. Δt is the time interval between consecutive measurement points or the simulation time step (sec). T is the total duration of the vehicle simulation (sec). Road Load Coefficients (F0, F1, and F2) have been derived from a database developed by the JRC during its vehicles measurement campaigns. More details can be found in (Pavlovic et al. 2016).

3.4 Simulation Scenarios

CAVs deployment in the market will be performed gradually. Therefore, in order to see the effect of the CAVs over a network, several simulations should be run with variable CAV penetration rates. In this work, a number of scenarios have been implemented with five different penetration rates 0, 25, 50, 75, and 100%. Moreover, traffic demand is not expected to remain the same over the CAV deployment period, as no the impact of CAVs on the overall demand remains to be shown. It is also interesting for the reader to see how variations on the demand can boost or hide the benefits from the new technology. So the same simulations were run, with traffic demand either 20% higher or lower than the calculated realistic one. In order to smoothly load and discharge the network, each simulation lasts 3 h, one hour with a lowered demand, one hour with the peak demand (main interest) and one last hour of again lowered demand to allow a realistic simulation of the traffic evolution.

4 Results

In this section, the main preliminary results of the present work are highlighted and discussed. The first observations refer to the time series of the average network speed. During the first and the last hours of the simulation, the network is loaded with reduced demands, which helps in realistic transition to the peak hour demand. When the network becomes congested (during the second hour), the vehicles' average speed drops in some scenarios up to 80% of the previously stabilized speed value. The graphs in this section present the aggregated average values (speed and deviation) in 10-minute periods.

In Fig. 3, it is observed that during the first hour of the simulation, the average speed for vehicles with CACC capability is lower than the one for those without CACC capability. This happens because CACC is not designed with tolerances and it limits maximum vehicle speed to the road limit.

During the second hour of the simulation, when the traffic demand is that of the peak hour, it can be observed that the higher the market penetration, the better the system response is. The high penetration rates manage to maintain the vehicles' average speed during peak hour to higher values, and the bounce back effect to the normal average network speed, when the traffic demand lowers, is performed much faster. Results demonstrate that the impact of connectivity and automation is small in case of uncongested networks. At the same time, they also show that the impact of the CACC technology in case of oversaturated networks is also much reduced, due to the limited capability of the reduced reaction time, to significantly increase the system capacity.

Since the impact of automated driving on the overall demand is not yet clear or predictable, it is worth investigating the effect of varying the overall traffic demand.

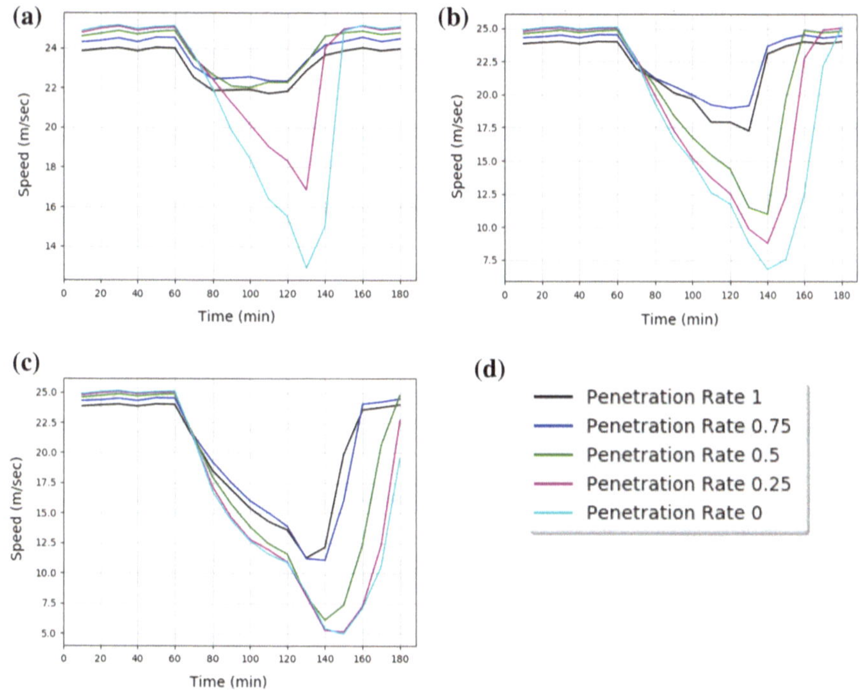

Fig. 3 Time series of the average network speed within the simulation period. Chart **a** refers to 80% of the actual traffic demand, **b** to 100% of the demand, and **c** to 120% of the actual demand, while chart **d** reports the figure legend

Consequently, it is interesting to observe that for 0% CACC penetration in Fig. 3a (80% of peak demand), the minimum average speed (approximately 13 m/s) is comparable to the minimum average speed at 50% penetration in Fig. 3b (100% of peak demand) and approximately at the minimum average speed at 100% penetration rate in Fig. 3c (120% of peak demand). In the present case, we can roughly estimate that 50% increase in CAVs penetration is equivalent to a reduction of 20% in traffic demand.

The time series of the speed standard deviation is the second metric used in the analysis. Observations previously outlined are also confirmed here. As soon as congestion appears, the standard deviation of the vehicles' average speed starts to increase as vehicles' behavior becomes unstable. While in the first two charts of Figs. 4, with 75 and 100% market penetration, the capacity of the network has not been reached, one can observe that for increased peak hour demand Fig. 4c, the demand overcomes the capacity. In scenarios with CACC penetration 0, 25, and 50% the traffic flow does not manage to become stable in the 3 h period of the simulation and only with penetrations 75 and 100%, the network quickly returns to a stable state. Smaller standard deviation of speeds contributes to safer and more comfortable road transport.

The time series of traffic flow and density are not presented here due to space limitation issues; however, it should be noted that they lead to similar conclusions.

4.1 Energy Consumption

The total energy requested by the vehicles' motion is calculated by integrating over time the power required at the wheels as given by Eq. 5. In order to compute the energy consumption over a network, we would need a distribution of the vehicle brands that fill the network, which is something that is not available to the authors and cannot be easily estimated. Therefore, we consider one vehicle as a representative vehicle for the entire network. It is worth mentioning that in this way, a quantitative analysis of the energy requested by the network cannot be derived in absolute terms, but in order to compare the results of different scenarios, the approach of considering a single representative vehicle can be considered sufficiently appropriate.

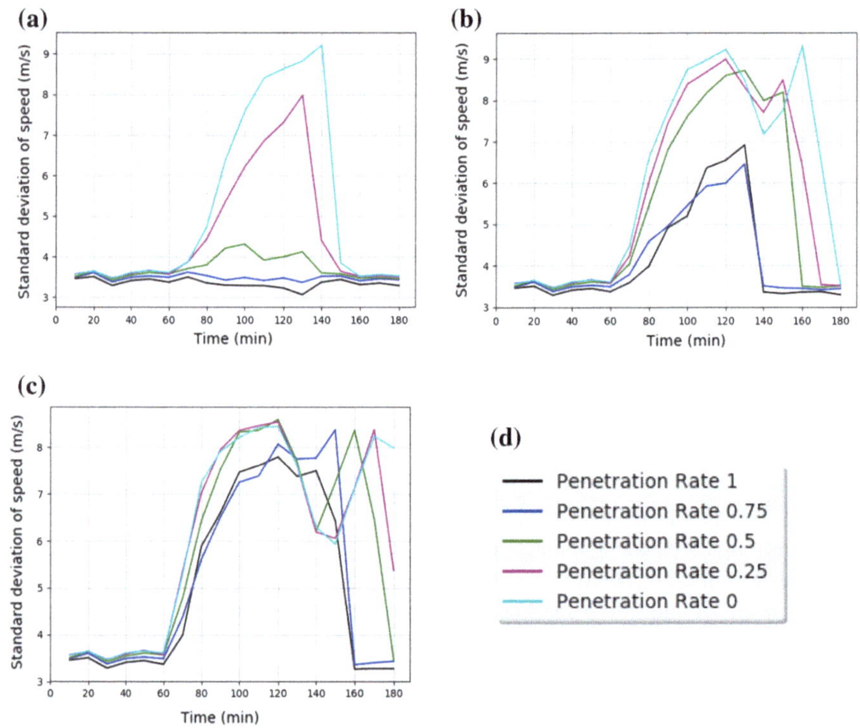

Fig. 4 Time series of the speed standard deviation within the simulation period. Chart **a** refers to 80% of the actual traffic demand, **b** to 100% of the demand, and **c** to 120% of the actual demand, while chart **d** reports the figure legend

Table 1 Average energy consumption of the network in kJ

CACC Penetration rate	Demand level		
	0.8*peak (kJ)	peak (kJ)	1.2*peak (kJ)
0	3468.9	3507.6	3539.5
0.25	3524.4	3579.9	3584.0
0.5	3602.6	3682.4	3703.9
0.75	3662.0	3838.3	3891.8
1	3618.1	3835.8	4070.8

As previously shown, higher market penetration rates of CACC increase the average vehicle speed. In this case, it is also expected that vehicle accelerations will be lower (less stop-and-go situation). However, the increase in the average speed seems, in this case, more important than the decrease in the acceleration and the overall energy consumptions seem to remain almost constant if not increasing. This result contradicts the popular notion that automated vehicles will be more energy efficient. Although it is true that energy efficiency does not always mean fuel efficiency (European Environment Agency 2017) and moreover the presented case study does not involve urban traffic, the present work gives a clear indication that CACC technology can help on reducing congestion, but it will not necessarily have the same effect on the reduction of energy consumption, fuel consumption and emissions, let alone that any effect positive or negative depends on traffic demand (Table 1).

5 Conclusions

In the present work, a microscopic traffic simulation framework for the simulation of normal and connected and automated vehicles (CAVs) is presented. The simulation framework has been developed for a real road network presenting heavy traffic dynamics, the ring road of Antwerp. Using the simulation, a number of scenarios have been tested with different CACC penetration rates and different traffic demand.

From the results, it is apparent that the use of CACC has substantial effect on the network performance. However, the effectiveness of such systems also depends on the overall traffic demand. In case CAVs will bring an increase in the overall demand for personal mobility, the promises they bring to reduce congestion and the other transport externalities will not necessarily hold. Furthermore, in case of any kind of communications breakdown, especially during rush hour, the increased demand with no connection capability can bring the system to collapse.

The presented results are motivating a series of future research works. It is the authors' opinion that with the use of V2I communication, different strategies can be implemented uniformly on the network or separately on each section to avoid the networks condition deterioration even for the higher traffic demands. Using the developed simulation framework it is possible to assess a plethora of different strategies and make useful notes for the inclusion of CAVs on the road networks.

References

Alonso Raposo M, Ciuffo B, Makridis M, Thiel C (to be published) The r-evolution of driving: from Connected Vehicles to Coordinated Automated Road Transport (C-ART). Eur Comm

Arnaout GM, Bowling S (2014) A progressive deployment strategy for cooperative adaptive cruise control to improve traffic dynamics. Int J Autom Comput 11:10–18. doi:10.1007/s11633-014-0760-2

Calvert SC, van den Broek THA, van Noort M (2012) Cooperative driving in mixed traffic networks - Optimizing for performance. In: 2012 IEEE Intelligent Vehicles Symposium. pp 861–866

Chin H, Okuda H, Tazaki Y, Suzuki T (2015) Model predictive cooperative cruise control in mixed traffic. In: IECON 2015—41st Annual Conference IEEE Industrial Electronics Society. pp 003199–003205

Degraeuwe B, Thunis P, Clappier A, Weiss M, Lefebvre W, Janssen S, Vranckx S (2016) Impact of passenger car NOx emissions and NO2 fractions on urban NO2 pollution—Scenario analysis for the city of Antwerp, Belgium. Atmos Environ 126:218–224. doi:10.1016/j.atmosenv.2015.11.042

Do lower speed limits on motorways reduce fuel consumption and pollutant emissions?—European Environment Agency. https://www.eea.europa.eu/themes/transport/speed-limits. Accessed 1 Jun 2017

Fakharian Qom S, Xiao Y, Hadi M (2016) Evaluation of Cooperative Adaptive Cruise Control (CACC) vehicles on managed lanes utilizing macroscopic and mesoscopic simulation

Frank M, Wolfe P (1956) An algorithm for quadratic programming. Nav Res Logist Q 3:95–110. doi:10.1002/nav.3800030109

Gipps PG (1981) A behavioural car-following model for computer simulation. Transp Res Part B Methodol 15:105–111. doi:10.1016/0191-2615(81)90037-0

Ilgin Guler S, Menendez M, Meier L (2014) Using connected vehicle technology to improve the efficiency of intersections. Transp Res Part C Emerg Technol 46:121–131. doi:10.1016/j.trc.2014.05.008

Kloostra B, Roorda MJ (2017) Fully autonomous vehicles: analyzing transportation network performance and operating scenarios in the Greater Toronto Area, Canada

Lefebvre W, Vercauteren J, Schrooten L, Janssen S, Degraeuwe B, Maenhaut W, de Vlieger I, Vankerkom J, Cosemans G, Mensink C, Veldeman N, Deutsch F, Van Looy S, Peelaerts W, Lefebre F (2011) Validation of the MIMOSA-AURORA-IFDM model chain for policy support: Modeling concentrations of elemental carbon in Flanders. Atmos Environ 45:6705–6713. doi:10.1016/j.atmosenv.2011.08.033

Lefebvre W, Van Poppel M, Maiheu B, Janssen S, Dons E (2013) Evaluation of the RIO-IFDM-street canyon model chain. Atmos Environ 77:325–337. doi:10.1016/j.atmosenv.2013.05.026

Litman T (2015) Autonomous vehicle implementation predictions: Implications for transport planning

Lu X-Y, Kan XD, Shladover SE, Wei D, Ferlis RA (2017) An enhanced microscopic traffic simulation model for application to connected automated vehicles

Mahmassani HS (2016) 50th anniversary invited article—autonomous vehicles and connected vehicle systems: Flow and operations considerations. Transp Sci 50:1140–1162. doi:10.1287/trsc.2016.0712

Ngoduy D (2013) Instability of cooperative adaptive cruise control traffic flow: A macroscopic approach. Commun Nonlinear Sci Numer Simul 18:2838–2851. doi:10.1016/j.cnsns.2013.02.007

Open street maps. https://planet.openstreetmap.org/. Accessed 31 May 2017

Pavlovic J, Marotta A, Ciuffo B (2016) CO2 emissions and energy demands of vehicles tested under the NEDC and the new WLTP type approval test procedures. Appl Energy 177:661–670. doi:10.1016/j.apenergy.2016.05.110

Shladover S, Su D, Lu X-Y (2012) Impacts of cooperative adaptive cruise control on freeway traffic flow. Transp Res Rec J Transp Res Board 2324:63–70. doi:10.3141/2324-08

van Arem B, van Driel CJG, Visser R (2006) The impact of cooperative adaptive cruise control on traffic-flow characteristics. IEEE Trans Intell Transp Syst 7:429–436. doi:10.1109/TITS.2006.884615

Xiao L, Wang M, van Arem B (2017) Realistic car-following models for microscopic simulation of adaptive and cooperative adaptive cruise control vehicles

Germany's New Road Traffic Law—Legal Risks and Ramifications for the Design of Human-Machine Interaction in Automated Vehicles

Christian Kessel and Benjamin von Bodungen

Abstract Germany has very recently adopted the allegedly most advanced road traffic law in the world. At first glance, drivers will be allowed to pursue non-driving-related activities during phases of automation such as surfing the internet or watching films. This paper shows, however, that the new law gives rise to significant legal uncertainty regarding human–machine interaction in automated vehicles, with liability risks emanating therefrom both for manufacturers and users of the new technology. This paper puts forward suggestions as to how these legal risks can be addressed when designing and implementing the human–machine interface. Automation systems will only find general acceptance if they can be brought to market without unduly burdening manufacturers and users from a liability perspective.

Keywords Automation · Civil liability · Constructional deficiencies · Federal road traffic act · Human-machine interface · Instructional errors · Product liability act · Vienna convention on road traffic

1 Introduction

On June 21 2017, several amendments to Germany`s Federal Road Traffic Act (*Straßenverkehrsgesetz*—hereinafter referred to as "StVG") entered into force that aim at creating legal certainty for manufacturers and users of automated vehicles. In particular, the legislative amendments establish rules for the interaction between vehicles equipped with automated driving functions and their users (Federal Government 2017).

C. Kessel
Bird & Bird LLP, Marienstraße 15, 60329 Frankfurt/Main, Germany
e-mail: christian.kessel@twobirds.com

B. von Bodungen (✉)
German Graduate School of Management and Law, Bildungscampus 2, 74076 Heilbronn, Germany
e-mail: benjamin.vonbodungen@ggs.de

C. Zachäus et al. (eds.), *Advanced Microsystems for Automotive Applications 2017*,
Lecture Notes in Mobility, DOI 10.1007/978-3-319-66972-4_19

This paper first gives a short overview of the very recent amendments to the StVG insofar as they relate to human–machine interaction in automated vehicles (Chap. 2). It then analyses their legal ramifications and shortcomings from the user's perspective (Chap. 3). Finally, the legal prerequisites for the design and implementation of human–machine interaction will be discussed against the background of the legislative amendments (Chap. 4). Manufacturers face substantial product liability claims if they violate these requirements. The paper also puts forward several suggestions as to how these legal risks can be mitigated.

2 The Amendments to the Federal Road Traffic Act

2.1 Levels of Automation Addressed

As indicated by the heading of the newly incorporated Section1a, the Federal Road Traffic Act now addresses highly and fully automated driving functions. It is to be noted, however, that the new statutory regime does not refer to the automation levels as defined by the Society of Automotive Engineers (SAE). Rather it builds on the nomenclature developed by the German Federal Highway Research Institute (*Bundesanstalt für Straßenwesen—BASt*) (Gasser et al. 2012). In essence, the term "high automation" in the German law is akin to SAE level 3 ("conditional automation"), while the term "full automation" equals SAE level 4 ("high automation").

A legal definition of vehicles with highly and fully automated driving functions is now contained in Section 1a (2) StVG. This provision lays down the requirements these systems must meet, namely that the automation system (1) takes over the vehicle control (including both steering and acceleration/deceleration) when activated, (2) observes all traffic rules during phases of automation, (3) can be overridden or switched off by the driver at any time, (4) recognizes when the driver has to drive the vehicle in person, (5) optically, acoustically, tactilely or by other means alerts the driver with adequate lead time if it is necessary to take over the driving functions, and (6) informs the driver that the automation system is being used contrary to the manufacturer's system description.

In summary, the statutory definition heavily rests upon the system descriptions developed by the Federal Highway Research Institute. In this way, the technical and legal terminology is synchronized to a large extent, which is to be welcomed. However, the Federal Road Traffic Act goes beyond the BASt-nomenclature by requiring that the automation system is capable of complying with all relevant traffic rules, and that the driver can override or switch off the system at any time. The latter criterion derives from, but is stricter than, the recently amended 1968 Vienna Convention on Road Traffic that also allows for the use of systems that cannot be overridden or switched off by the driver, insofar as they are in conformity with other international legal instruments.

If any of the abovementioned requirements contained in Section 1a (2) StVG is not met by an automated driving function, the newly enacted rules for the use of highly and fully automated driving systems described subsequently will not apply.

2.2 *Definition of "Driver"*

Section 1a (4) StVG provides that the "driver" is not only the person steering the vehicle but also the person who has activated a highly or fully automated driving function, even if he or she is not personally steering the vehicle. By means of this provision, the German legislator wanted to take into account scholarly opinions in German legal literature that doubted whether the user of a highly or fully automated vehicle is still to be considered as a "driver" for the purposes of the Federal Road Traffic Act (von Bodungen and Hoffmann 2016). The legislator also wanted to stress the importance of the presence of a driver, so that German law so far does not permit autonomous driving (where there may be no person at all to intervene in the driving process).

2.3 *Interaction Between the Automation System and the Driver*

The newly incorporated Section 1a (1) StVG stipulates that the operation of a vehicle by means of a highly or fully automated driving function is permissible if this function is used by the driver in accordance with its intended use. For example, an automation system designed exclusively for motorways must not be used in inner-city traffic. Conversely, the legislative definition outlined above requires highly and fully automated systems to alert the driver that he or she is using the system in an improper manner.

At the heart of the latest amendments to the Federal Road Traffic Act lies the newly introduced Section 1b which allows the driver to turn away from both the traffic situation and the vehicle control while the highly or fully automated system is activated. However, the system user must at any time be ready to take over control of the vehicle without undue delay after the system has made him or her aware of the necessity to resume the manual steering function. As outlined above, highly and fully automated systems by definition must express such take over requests with an adequate time reserve, and they may not surrender vehicle control to their human user without any lead time. Furthermore, the system user must also resume the driving task when he or she realizes, or due to obvious reasons must realize, that the conditions for the proper use of the automated system no longer are existent. This presumably would be the case if the driver took notice of the siren of an approaching police car because automation systems are currently not advanced enough to perceive or even distinguish sounds.

In sum, the legislative amendments establish a set of rules for the interaction between automated driving functions and their users. In particular, the mandatory system features and the system user's rights and obligations are closely interlinked and work together. Although the statutory language seems rather comprehensive at first glance, this paper will subsequently look into the question of whether the outlined amendments to the Federal Road Traffic Act indeed accomplish the aim of establishing legal certainty for manufacturers and users of such automation systems in Germany.

3 The Statutory Amendments from the Driver's Perspective

3.1 Brief Overview of the Statutory Liability Regime for Drivers

Traditionally, German law has always placed drivers under fault-based liability. Such fault is presumed under Section 18 (1) StVG in case of a road accident. It is then up to the driver to provide evidence that he or she was driving in accordance with the road rules and did not negligently or even intentionally cause the personal injury or property damage at issue. The counterpart to this presumption of fault is the statutory stipulation of a maximum liability amount. If the damage is incurred due to the use of a highly or fully automated system, liability is capped by the recent legislative amendments at 10 million Euros for personal injury and two million Euros for property damage. This is equivalent to a doubling of the liability limits for damages caused using conventional vehicles.

In addition, drivers are subject to unlimited liability under German tort law. Again, such liability rests upon culpable conduct on the part of the driver. However, tort law places the burden of proof of such culpable behavior on the injured party seeking compensation.

In sum, the recent legislative amendments do not unsettle the fundamentals of the civil liability regime for drivers in Germany. The only substantial modification is the increase of the liability limit mentioned above.

3.2 Ramifications of the Obligations Imposed on Automated System Users

As described above, the modified Federal Road Traffic Act establishes several new behavioral obligations that must be observed by the users of automated driving systems. If the system user breaches any of these obligations and such breach occurs at least negligently, then he or she will be liable if a third party's life, limb, or property is damaged as a consequence.

3.2.1 Obligation to Use the Automation System Properly

First of all, the statutory obligation to use the automation function only for its intended purpose is a source of potential liability. If the driver disregards the limited range of application of a highway pilot and uses such automation system in urban traffic, he or she obviously will be liable in case an accident occurs.

However, the allocation of liability will not be as clear as this in all cases. This particularly concerns the fact that highly and fully automated systems so far are unknown to the market. In addition, the exact scope of their performance may differ from system to system. Consequently, the system user will need to very carefully acquaint itself with the limits of the system employed and the exact functioning of the human-machine interface. If he or she neglects to do so, such behavior is likely to result in civil liability in the case of damage caused to a third party (von Bodungen and Hoffmann 2016). Against this background, the manufacturer's system description gains pivotal importance for the information of the system user. In essence, it shapes the standard of due care the system user must satisfy to escape liability. It is also to the benefit of the driver that the system must call his or her attention to an improper use of the automation function. In this way, the driver is enabled to terminate its misuse of the system and avoid liability.

3.2.2 Sharing of the Driving Task Between the Driver and the Automation System

At first glance, the main achievement of the legislative amendments would seem to be that during phases of high and full automation the system user may engage in non-driving related activities. Yet, the new law also establishes the requirement that the system user must remain vigilant in order to resume the driving task "without undue delay" after a take over request from the system.

The well-established legal concept of undue delay implies that the system user must resume control as soon as this is possible and reasonable in the present circumstances. However, in the absence of any guidance from the legislator, there remains a large degree of legal uncertainty in this regard. In particular, the recent statutory amendments do not touch upon the findings of modern traffic psychology in relation to the limitations of the human performance at the human–machine interface when resuming control from an automated system. For instance, it has been shown that higher degrees of automation results in a fragmented state of knowledge in relation to the specific traffic situation as well as extended reaction and response time on the part of the human driver (Fastenmeier and Schlag 2016). There are studies that surmise a transition period of up to 40 s before the driver has regained sufficient and stable control of the vehicle from the automated system (Merat et al. 2014). It is completely unclear whether the courts will indeed grant drivers such a long transition period when assessing liability.

The legal uncertainty inherent in the statutory amendments is increased by the fact that the legislator envisages situations where the driver must intervene and take over control without a prior system warning. This is less of a problem when the driver has taken note of the fact that the conditions for the proper use of the automation system no longer exist. However, the Federal Road Traffic Act also requires intervention when such circumstances are obvious, irrespective of whether or not the driver has actually taken note thereof. The legislative explanatory memorandum refers to technical and other malfunctions in the drive mode such as a tire blow-out (Federal Government 2017). Beyond these obvious cases, the law remains silent and leaves it to the courts to flesh out its vague terms. Prior to the existence of settled case law, users of highly and fully automated systems thus face additional uncertainty as to their behavioral obligations in the context of human–machine interaction.

Contrary to the announcement by the Federal Government (Federal Government 2017), it seems doubtful whether system users will indeed be able to relax and turn away from the driving task for the purpose of engaging in other activities. In the absence of clear legislative language to this effect, any driver risks not reacting quickly enough to a system request in the eyes of a court. From the driver's perspective, it would also be desirable if future legislation spelled out at least some of the non-driving related activities permitted during phases of automation. Such activities the driver could pursue without risking liability claims for not having been sufficiently vigilant could include checking emails, surfing the internet, or streaming films. Due to the current lack of such legal clarification, the risk of running into liability claims with open eyes essentially forces the driver to monitor both the automation system and the traffic at all times. This obviously greatly diminishes the utility of automated functions. The new law has therefore already been criticized by other commentators for being too restrictive to be useful in practice (Schirmer 2017).

4 Liability Issues from the Manufacturer's Perspective

4.1 Brief Overview of the Statutory Liability Regime for Manufacturers

If a defective product that has been marketed in the European Economic Area causes damage to consumers, the manufacturer has to provide compensation irrespective of whether or not there is negligence or fault on its part. This is set forth in the EU Product Liability Directive 85/374/EEC which has been implemented in Germany by way of the Product Liability Act (*Produkthaftungsgesetz*). A product such as an automated vehicle or component thereof will be deemed defective if it does not provide the safety which its user is entitled to expect, taking all circumstances into account, including the presentation of the product, the use to which it could reasonably be expected that the product would be put, as well as the time when the product was put into circulation. In addition, product liability can arise under national tort law. As described above, this liability regime is fault-based and applies only if the manufacturer at least negligently breached its duty of reasonable care and such breach caused damage to the claimant's life, limb, or property.

Both liability regimes apply to end manufacturers as well as producers of defective component parts. The two liability systems largely run in parallel if the design of a product, its fabrication, or the manufacturer's product-related instructions did not meet the injured person's legitimate safety expectations. The greater the dangers are that emanate from a product, the higher the standard of care is that the courts require manufacturers to observe when bringing their products to market.

4.2 Product Liability Issues in Relation to Automated Vehicles

Taking into account the guiding principles of the German product liability regime, two sources of liability seem particularly relevant for manufacturers in relation to automated vehicles: constructional deficiencies and instructional errors.

4.2.1 Constructional Deficiencies

As described above, the definition of highly and fully automated systems in Section 1a (2) of the amended Federal Road Traffic Act lists several constructional requirements. In particular, manufacturers must ensure that highly and fully automated systems are capable of observing the rules of the road such as speed limits or priority rules, thus protecting the users of the vehicle, the users of other vehicles participating in road traffic and/or innocent bystanders. Furthermore, automated systems must recognize the need to retransfer vehicle control to the driver and alert him or her thereof with adequate lead time. Consumers' legitimate safety expectations are shaped by these legal stipulations. If the design of a vehicle falls short thereof, this will trigger product liability in the event of damage. By way of an example, this would be the case if the system unexpectedly surrendered vehicle control to the driver while he or she was not anticipating such instant hand over at the human-machine interface.

Beyond the constructional design features prescribed by the amended Federal Road Traffic Act, currently no clear legal guidance is available as to the exact level of safety consumers may reasonably expect from automated vehicles. Yet, manufacturers must bear in mind that road accidents pose a threat to human life and health and thus the courts will presumably apply a stricter approach when passing the judgement on whether or not the design of an automation system lived up to the legitimate safety expectations of its user, other motorists and/or innocent bystanders. Consequently, there is much to suggest that automated vehicles must have self-surveillance features regarding their condition in order to detect potential system failures in time.

Next, product liability law requires manufacturers to consider foreseeable product misuse already at the design stage. Behavioral psychology has proven a significant decline of human vigilance when transitioning from manual vehicle control to the rather monotonous supervision of an automation system (Bengler 2015). It thus seems highly likely that drivers may become drowsy and take a nap during phases of automation. Due to this misuse of the automation function, they may miss a system request to resume the driving task and thus cause an accident. Product liability law expects manufacturers to take constructional counter-measures to avoid such improper use of the system. This could be achieved by way of eye-tracking technology and the activation of warning tones and signals if the system detects that the driver is about to fall asleep (Winner et al. 2015).

Finally, automation systems are not yet established in the market, and therefore system users will initially not be acquainted with how they should best act at the human–machine interface. From a product liability perspective, it would seem imperative that manufacturers pay special attention to the design of the display and operating elements of automation systems. The design should be as intuitive and user-friendly as possible. This is particularly true in light of the fact that changing mobility habits may entail that drivers no longer own their cars but rather avail themselves of shared mobility solutions. If this holds true, drivers will be using different vehicle models far more often than this has traditionally been the case. Therefore, manufacturers should desist from designing model-specific—isolated—solutions for the human–machine interface as this would presumably collide with the system users' justified expectation of safely being able to use different vehicle models without the necessity of major behavioral adjustments.

4.2.2 Instructional Errors

Due to the novelty of automated vehicles, manufacturers will have to pay special attention to satisfying their instructional obligations. An instructional error will be assumed by the courts if the manufacturer omits to warn users of product risks that cannot be mitigated by constructional means.

To avoid product users suffering damages from the proper use—or the foreseeable misuse—of an automated vehicle, manufacturers should advise customers very accurately, comprehensibly, and intuitively about the operational purpose, performance and limitations of the relevant automated system. Special attention must thereby be paid to the allocation of responsibilities at the human–machine interface (von Bodungen and Hoffmann 2016). The manufacturer best knows the safety-relevant features of its product and must fully inform its customers thereof.

For instance, if the manufacturer's system description remains vague, incomplete or even incorrect with respect to the circumstances when the driver must resume the driving task, this can result in liability on the part of the manufacturer due to an instructional error. The manufacturer's system description will have to very precisely define and distinguish the responsibilities of the human driver and the automated system. In order to satisfy their instructional obligations, manufacturers should consider playing in relevant information on the proper system use via the on-board system. It also could be advisable for manufacturers to offer a driver's training on the occasion of the purchase of a vehicle.

Finally, the amendments to the Federal Road Traffic Act described above require automation functions to warn drivers in case of improper system use. Such instruction will only live up to justified consumer safety expectations if it puts the driver in a position to react adequately in precarious traffic situations.

5 Summary

The recent amendments to the Federal Road Traffic Act aim at positioning the Federal Republic of Germany as the legal pioneer for automated driving systems. By creating legal certainty for the operation of highly and fully automated systems, the German Federal Government strives to promote investment safety in Germany for manufacturers in the automotive industry.

In contrast, this paper has demonstrated that the recent amendments to the Federal Road Traffic Act are far from clear. Notably the rules of conduct imposed on users of automated systems are formulated rather vaguely. This leaves much room for uncertainty unless and until the courts address and adjudicate these issues. The ongoing standardization particularly undertaken at the international level by the United Nations Economic Commission for Europe (UNECE) may prove helpful in this regard.

For the moment, however, the risk of running into liability essentially forces drivers to shy away from non-driving related tasks and remain vigilant at all times, which undermines the anticipated advantages of automation. Special attention will have to be paid to the design and operation of the human–machine interface. Drivers so far using manually controlled—conventional—vehicles will have to come to grips with completely new vehicle features. From a liability perspective, employing automation systems requires users to familiarize themselves very comprehensively with the functioning and limitations of the automation functions at hand. At the same time, their legitimate safety expectations place significant burdens on manufacturers. Above and beyond the design specifications prescribed in the recent amendments to the Federal Road Traffic Act, manufacturers will have to pay special attention to appropriate system design at the human–machine interface. In addition, adequate consumer information will experience a major increase in importance. Legal uncertainties remain in this area as there is no guidance as to the required format, level of detail and degree of comprehensibility of such information. Sensibly, the legislator has already announced that a review of the application of the new legislative provisions will be carried out in 2020. It remains to be seen how extensively the legal framework for automated driving in Germany will have to be revised thereafter.

References

Bengler K (2015) Grundlegende Zusammenhänge von Automatisierung und Fahrerleistung. Zeitschrift für Verkehrssicherheit 61:169–173

Fastenmeier W, Schlag B et al (2016) Hochautomatisiertes oder autonomes Fahren als wünschenswerte Zukunftsvision? Offene Fragen mit Blick auf die Mensch-Maschine-Interaktion, DGVP Infos—Positionen—Empfehlungen 03/2016

Federal government, bulletin no. 38-3 dated 30 March 2017

Federal Government (2017) Explanatory memorandum to the 8th amendment act to the federal road traffic act, Bundestag Document 18/11300

Gasser T, Arzt C et al (2012) Rechtsfolgen zunehmender Fahrzeugautomatisierung—Gemeinsamer Schlussbericht der BASt-Projektgruppe (BASt-Bericht F 83). Wirtschaftsverlag NW, Bremerhaven

Merat N, Jamson A et al (2014) Transition to manual: driver behaviour when resuming control from a highly automated vehicle. Transp Res Part F 27:274–282

Schirmer J (2017) Augen auf beim automatisierten Fahren! Die StVG-Novelle ist ein Montagsstück, Neue Zeitschrift für Verkehrsrecht 30:253–257

von Bodungen B, Hoffmann M, (2016) Autonomes Fahren - Haftungsverschiebung entlang der supply chain? (1. Teil), Neue Zeitschrift für Verkehrsrecht, vol 29, pp 449–454

von Bodungen B, Hoffmann M (2016) Autonomes Fahren - Haftungsverschiebung entlang der supply chain? (2. Teil), Neue Zeitschrift für Verkehrsrecht, vol 29, pp 503–509

Winner H, Hakuli S et al (2015) Handbuch Fahrerassistenzsysteme, 3rd edn. Springer Vieweg, Wiesbaden

Losing a Private Sphere? A Glance on the User Perspective on Privacy in Connected Cars

Jonas Walter and Bettina Abendroth

Abstract Connectivity is one of the major prerequisites of automated driving. Enabled by numerous connected sensors, new cars offer new functionalities, provide higher security levels and promise to enhance the comfort of travelling. However, by connecting a vehicle with its environment, the car becomes more transparent. The integration of the car into a smart grid seems to conflict with the users' expectation of their car as a private retreat, thus reducing the acceptance and usage adoption of connected cars. This article aims at helping developers and engineers to consider the user's expectations when designing a connected car. Furthermore, this article reviews and compares recent international surveys on user's privacy with our own results on the user's attitude towards connected vehicular services.

Keywords Connected car · Privacy · User · Acceptance

1 Introduction

Modern cars offer high levels of safety and comfort. An increasing degree of automation constantly extends the car's ability to anticipate the current traffic situation, thus reducing the workload which is imposed on the driver (De Winter et al. 2014). For example, advanced driver assistance systems guide the driver's behavior (Dotzauer et al. 2015) or detect and communicate potential crashes (Dokic et al. 2015) in order to prevent risky manoeuvers. Moreover, future cars are likely to take over all driving tasks as car manufacturers pursue high automation levels (Chen and Englund 2016). However, high automation levels heavily rely on so-called vehicular ad hoc networks, which transfer data between multiple entities like cars or

J. Walter (✉) · B. Abendroth
Institute of Ergonomics and Human Factors, Technische Universität Darmstadt,
Otto-Berndt-Straße 2, 64287 Darmstadt, Germany
e-mail: j.walter@iad.tu-darmstadt.de

B. Abendroth
e-mail: abendroth@iad.tu-darmstadt.de

C. Zachäus et al. (eds.), *Advanced Microsystems for Automotive Applications 2017*,
Lecture Notes in Mobility, DOI 10.1007/978-3-319-66972-4_20

infrastructure (Papadimitratos et al. 2009). Hence, a constantly growing number of sensors observe and communicate the environment as well as the interior including the passengers. While the advent of these means of vehicular ambient intelligence fosters new functionalities, provides higher security levels, and promises to enhance the comfort of travelling (Steg 2009), it might change the perception of our car as private which has been persisting so far.

Private cars are more than a simple means of transport. Car use does not only fulfil pragmatic functions like getting from A to B, but also has affective and symbolic significance (Sheller 2004). Among others, driving our private car can be associated with pride and enjoyment. People build a relationship towards their vehicles and from an emotional bond to their car (Gardner and Abraham 2007). This connection underpins the pursuit of safety, enjoyment, and autonomy, but also the desire for a private refuge (Huang et al. 2016). When commuting back home, we acknowledge our retreat and enjoy the car as private space. So far, cars have offered this private retreat that protected us from unwanted intrusions. However, with the integration of the car in the smart grid, the image of the car as a private refuge begins to crack. It is not a physical intrusion, which might compromise the driver's private sphere within a car, but a digital one. A multitude of sensors turns the car into a context-aware smart vehicle that relies on ubiquitous computing (Pohl et al. 2007). Sensors, most of which the driver is not aware of, not only detect the environment (Li et al. 2011; Benedettini et al. 2009) or the current state of single components (Rebolledo-Mendez et al. 2014), but also sense the driver's health and cognitive state (Chui et al. 2016; https://www.webofknowledge.com/2017. This poses the question, if the connected car can still be a private place even if it registers and communicates processes that take place in the interior. There is no doubt that connectivity presents new options and enables extended functions, but how do drivers perceive the advent of connectivity within the car? While technological limits of smart cars are currently shifting forward, so far relatively little attention has been payed to the user's acceptance and expectations of connected cars. Therefore, to provide an overview of user acceptance of connected vehicular services, a systematic literature review of user studies on connected cars is conducted. Subsequently, we complement the results of the review with insights from our own user study. Finally, we derive practical implications.

2 Literature Review

2.1 Methodology

In order to gain an overview of the consisting literature on privacy-related user studies within the field of the connected vehicle, a systematic literature research was conducted. Using the Web of Knowledge database (Schoettle and Sivak 2014) the existing peer-reviewed literature was scanned using the key words "privacy vehicle", "privacy car", "connected car", "connected privacy" and "connected vehicle". Due to a large amount of hits (>3000), the last key word was refined using "user", making the final key word combination "connected vehicle user". For each search, a fixed selection process was administered. The first step was to check if each result dealt with a topic directly related to the connected car and if it reported any privacy-related user study. Only papers that reported a user study within the topic of privacy in the connected car were considered. Thus, neither technical papers on connected car technologies nor user studies on connected devices different from the vehicle were taken into account. Applying this scheme resulted in only five remaining papers. Table 1 reflects the number of hits for each individual search combination and categorizes the final papers. As reflected at the end of Table 1, an additional paper was added to the list of reported papers. Schoettle and Sivak (2014) was not found using Web of Knowledge. However, it is nonetheless a very valuable contribution to the elucidation of the user acceptance of connected cars and is therefore included here.

Thus, this article reviews five papers (published between 2012 and 2016). To categorize these papers with respect to their focus, several content focus are identified. First, Schoettle and Sivak (2014) provide a good overview within a comprehensive survey on conducted cars. These findings are smoothly complemented by Endo et al. (2016) with an overview of privacy setting acceptance in connected cars. Second, Eyssartier (2015), Eriksson and Bjørnskau (2012) put their focus on the acceptance of event data recorders and related technologies. Finally, Derikx et al. (2016) shed a light on the relevance of single attributes of connected services for users.

2.2 *Relevant Privacy Factors for the Adoption of Connected Services*

Connected vehicular services are trending, but are still uncommon. Users therefore have none or only little practical experience with these services. Do users even

Table 1 Search scheme and result categorization for literature review

Key word combination	Hits (unselected)	Identified paper
"privacy vehicle"	500	Eyssartier (2015), Eriksson and Bjørnskau (2012)
"privacy car"	89	Endo et al. (2016), Derikx et al. (2016)
"connected car"	54	–
"connected privacy"	526	–
"connected vehicle user"	157	–
@@*Additionally added (found at IEEE Xplore)*	–	Schoettle and Sivak (2014)

know about their existence? Under which circumstances would users use these services? To tackle these questions, Schoettle and Sivak (2014) conducted an international online survey with 1596 respondents from the United States, the United Kingdom, and Australia on connected cars in 2014. Most participants were unaware of connected cars. However, the authors reported a generally positive impression of connected cars. Accordingly, the participants expected multiple benefits of connected cars. They expected the number and severity of crashes to decline the strongest, while effects on traffic conditions and driver distraction were rated as least expected. On the other side, there were also manifold concerns, though the participants' concerns were less pronounced than the expected benefits. Exaggerated trust in the system, system failure, and legal liability issues were the strongest concerns, but participants also mentioned data privacy concerns. 69.3% of participants were concerned about data privacy in cars. Moreover, among connected car features, safety was the most important, followed by mobility and environment. Internet connectivity and the possibility to integrate one's smartphone in the car were moderately important. Finally, participants were moderately interested in connected vehicular technologies and were willing to pay on average 44$ extra for those technologies.

As for most technology adoptions, the acceptance of connected vehicular services is at least partly based on the weighing of expected benefits against anticipated costs (Dinev and Hart 2006). Anticipating this process, Endo et al. (2016) further elucidated relevant factors for the user acceptance of connected cars by conducting a qualitative interview study in Japan. To identify privacy concerns and factors engaging data disclosure, the authors presented 20 participants with 14 use cases of data utilization in connected vehicles. For each use case, the authors varied the number of involved parties in data processing. Together, the use cases covered safety and security functions as well as entertainment services. Endo et al. (2016) reported that participants rated services offering safety and security benefits more positively than infotainment-related services. Moreover, participants who drove alone were more accepting to some infotainment-related services like navigation assists, than those who had a co-driver. Critically, the range of data sharing as well as the nature of acquired data influenced the initial acceptance of a service. The more sensitive the data and the higher the number of data-receiving parties, the lower the acceptance of a service was. Subsequent in-depth interviews further highlighted the relevance of data parsimony. Participants felt uncomfortable in cases of intensive data consumption, but were quite easily encouraged to disclose data by transparent usage communication.

As most personal decisions, privacy-relevant decisions also have a social aspect (Eyssartier 2015). To study the influence of the social context, Eyssartier (2015) studied the acceptability of event data recorder (EDR) by conducting focus groups in two French civil services which agreed to the adoption of EDRs in their vehicles ($n = 28$). The authors found that social context is an important predictor of acceptability of EDR systems. While most respondents accepted an implementation of EDR in their professional vehicles, most refused to equip their private car as they perceived this to be an invasion of privacy. Moreover, the respondent's decision on

EDR acceptance and thus data disclosure depended greatly on data access, data usage and data identity. Next to the drivers, the authors identified manufactures as authorized parties to access the EDR data for the sake of vehicle safety. However, the systems released only car-related and driving behavior-related data, while retaining personal data allowing a direct identification (like video sequences). In contrast, management or insurance companies were completely excluded from data access.

Eriksson and Bjørnskau (2012) also picked up EDR technology, but compared it to further traffic safety measures with potential impact on privacy. In a large questionnaire study in Norway, Sweden and Denmark, the authors compared section control, informative intelligent speed adaption (ISA) and EDR ($n = 1319$). While the function of all techniques is related to crash prevention or crash investigation, the amount and frequency of data recording as well as transparency varies. While EDR is least transparent and might record the largest amount of data, section control is a modern technique for civil speed control that collects fewer data, as it is a stationary setup at certain roadsides. ISA, a driver assistant system dedicated to the support of speed control, is likely to collect more data than section control units but bears a higher transparency due to its immediate user feedback. Eriksson and Bjørnskau (2012) found that acceptability of the traffic safety measures varied with perceived privacy threat. Respondents perceived privacy infringement to be the highest in EDR, which thus had the lowest acceptability. Highest acceptability was indicated for section control. Moreover, the more transparent the measures were the higher acceptability was, as ISA was preferred over EDR.

The above-reported studies elucidated general acceptance of connected vehicular technologies. They demonstrated that privacy plays an important role when it comes to acceptability of new technologies. However, while the studies research the adoption of new technologies, they did not simulate real choice behavior. In contrast, Derikx et al. (2016) mimicked real choice behavior by choice-based conjoint analysis on mobile insurances. 60 participants indicated their preferences for various insurances that varied in the amount of data collection and consumer saving. Insurances were able to collect location and/or road behavior data. The insurances either kept this data, exploited it for additional offerings or forwarded it to third parties for advertisement purposes. The authors found that monetary savings could compensate privacy concerns. Specifically, privacy of behavior was more important to respondents than privacy of location. In contrast, participants declined data sharing with third parties. Hence, even if choice behavior is experimentally mimicked, data type, data receiver and the perceived strength of the benefit are important factors for privacy-relevant product decisions.

Taken together, our review yielded a limited number of studies that provided first insights on the impact of privacy characteristics on the acceptance of connected vehicular services. Though quite unknown to users so far, connected vehicular services are perceived as somewhat promising upcoming techniques, which are associated with distinct privacy concerns. All the studies identified the data type, the identity of the data receiver and the expected benefit as decisive privacy-relevant factors. Social context was found to be influential for adoption decisions as certain privacy-invasive services are accepted in professional, but not in private contexts. However, the studies show that transparency in data consumption and data procession as well as monetary incentives can compensate privacy concerns. Hence, these results provide us with a broad overview over relevant settings. However, to derive distinct practical implications, these insights are too shallow. For example, if the type of collected data is relevant for the adoption for a service, which data types do users perceive as sensitive? Furthermore, if users differentiate among data receivers, who do they trust? To gain more detailed insights in the user's opinion on connected vehicular services, we conducted an online survey study that tackled the following research questions (for more results see Müller et al. 2017):

- How relevant is privacy in comparison to safety, security or entertainment benefits?
- Which data in connected cars is sensitive to users?
- To whom would users release their data?
- Under which circumstances are users prepared to disclose their data?

3 User Study

101 participants (33 women) participated in an online survey on connected vehicular services. The mean age was 36.74 years (sd = 14,17). 87 participants (86,1%) possessed their own car. The majority of participants drove between 5000 and 15000 km per year (54,5%). All reported results stem from either tests for binomial distribution or one-sample student t-tests with $\alpha = 0.05$. If not reported differently, one-sample t-tests were tested against deviation from the center of the scale.

3.1 Results

First, we wanted to know if users perceived privacy to be more critical in cars than in other connected devices like smartphones. Participants did not report any differences (binary response options; 61.4% do not see any differences, test for binomial distribution: $p \leq 0.05$). Second, we attempted to locate the relevance of privacy in comparison with different types of benefits. For this purpose, we asked how much users would agree to release personal data in exchange for mobility, safety or comfort benefits. Participants indicated their agreement on a five-point Likert scale from "agree" (1) to "do not agree" (5). Users were ready to disclose personal data like their location for advanced traffic information in real-time ($M = 1.74$, $sd = 1.13$, $t(100) = -11.20$, $p \leq 0.001$) as well as for an automatic emergency call system (eCall; $M = 1.56$, $sd = 0.98$, $t(100) = -14.73$, $p \leq 0.001$). In contrast, users were undecided whether they should release their data for automatic hotel reservations at their travel destination (M = 3.24, sd = 1.42, t(100) = −1.68, p = 0.24).

Next, we identified the data types that the users perceived as personal. For this, participants rated their agreement to the sensitivity of a list of data on the Likert scale mentioned above. Table 2 lists the results for the rating. Most data belonging to the category of user preferences (like seat adjustment settings) were perceived as being personal as well as physiological, driving behavior related (like distance to the vehicle in front) and location data (all $p \leq 0.001$). An exception was air conditioning usage which was not found to be sensitive (M = 2.71, sd = 1.26, t(100) = −2.25, p = 0.027). Users only rated environmental data (like temperature) or operational characteristics (like engine temperature) as not being sensitive. In contrast, users rated data related to car usage differentially. While mileage was found to be sensitive (M = 2.33, sd = 1.19, t(100) = −5.60, $p \leq 0.001$), participants did not agree on the sensitivity of fuel consumption (M = 2.69, sd = 1.40, t(100) = −2.25, p = 0.029).

Aside from the data type, users see the identity of the data receiver as an important privacy factor. To elucidate whom users trust, we let our participants indicate their agreement on the trustworthiness of various parties on the same Likert scale as mentioned before. The parties the participants had to evaluate ranged from close relatives to providers of connected services. Participants had a strong trust in ambulance (M = 1.85, sd = 0.95, t(100) = −12.04, $p \leq 0.001$) and police

Table 2 Sensitivity of various data types (Likert scale from 1: *agree* to 5: *do not agree*)

Data type	M	sd	t	p
Contacts	1.65	1.00	−16.11	≤0.001
Music favorites	1.65	1.05	−12.88	≤0.001
Heart rate	1.70	1.11	−11.67	≤0.001
Location	1.97	1.14	−9.02	≤0.001
Seat adjustment	2.11	1.28	−6.93	≤0.001
Mean velocity	2.15	1.27	−6.67	≤0.001
Distance to lead vehicle	2.33	1.33	−4.98	≤0.001
Mileage	2.33	1.19	−5.60	≤0.001
Service interval	2.37	1.26	−5.00	≤0.001
Fuel consumption	2.69	1.40	−2.22	n.s.; 0.029
Air conditioning	2.71	1.26	−2.25	n.s.; 0.027
Engine temperature	3.65	1.36	4.70	n.s.; $t > 0$
Tire pressure	3.67	1.43	4.65	n.s.; $t > 0$
Environmental temperature	3.96	1.43	7.41	n.s.; $t > 0$

Note N = 101; one-sided t-test for X < 3 (α = 0.025). Sorted by perceived sensibility (M)

Table 3 Trust in data-receiving parties (Likert scale from 1: *agree* to 5: *do not agree*)

Data-receiving party	*M*	*sd*	*t*	*p*
Ambulance	1.85	0.95	−12.04	≤0.001
Police	2.16	1.24	−6.71	≤0.001
Family	2.40	1.30	−4.52	≤0.001
Traffic control center	3.00	1.36	0.00	>0.5
Breakdown service	3.28	1.27	2.15	≤0.05
Garage	3.49	1.29	3.82	≤0.001
Car manufacturer	3.59	1.29	4.53	≤0.001
Insurance	3.79	1.30	6.05	≤0.001
App provider	4.63	0.71	22.84	≤0.001

Note N = 101; Sorted by mean trust (M)

($M = 2.16$, $sd = 1.24$, $t(100) = -6.71$, $p \leq 0.001$), which was even stronger than for their own family ($M = 2.40$, $sd = 1.31$, $t(100) = -4.52$, $p \leq 0.001$). On the contrary, insurances ($M = 3.79$, $sd = 1.30$, $t(100) = 6.05$, $p \leq 0.001$) and app providers ($M = 4.63$, $sd = 0.71$, $t(100) = 22.84$, $p \leq 0.001$) were the least trusted parties. Table 3 displays the complete ratings for (dis-)trusted parties.

Finally, we wanted to identify a compensation threshold for data disclosure. As reported above, we found that perceived benefits in mobility and safety are strong motivators to disclose one's data. To highlight further incentives for disclosure, we confronted participants with certain circumstances and asked if they would agree to release their data. On the before mentioned Likert scale, participants indicated that they were not ready to release neither their car's operational characteristics ($M = 3.43$, $sd = 1.53$, $t(100) = 2.78$, $p \leq 0.01$) nor their driving profile ($M = 3.40$, $sd = 1.49$, $t(100) = 2.66$, $p \leq 0.01$) for monetary incentives.

3.2 Discussion

Our study contributed more detailed insights into the user's preference for privacy-relevant factors. While previous studies highlighted the relevance of data type and identity of the data-receiving party, we went a step further and elucidated which data are personal to users and who they trust. According to our results, users seem to be critical towards those parties that act on the private market. This even holds true for garages and breakdown providers. Moreover, our participants rated a broad range of data to be sensitive, while seemingly uncritical data like tire pressure has been shown to bear the potential to reveal delicate information about the passengers (Jensen et al. 2016). Surprisingly, we could not find a positive effect of monetary incentives on data disclosure. Hence, our results are in contrasts to those of Derikx et al. (2016). However, as Derikx et al. (2016) assessed the effect of monetary incentives in a more concrete manner, methodical differences might have caused our diverging results.

4 Conclusion and Practical Implications

The advent of connected cars offers plenty of new technological options, most importantly automated driving (Chen and Englund 2016). However, connecting the car with its environment means transferring a lot of data which might include data directly related to the passengers. Hence, connected cars are relevant to the passenger's privacy. Therefore, this paper raised the question if the loss of a private zone is associated with the connectivity of the car.

As our literature review demonstrated, the connected car has so far been mainly researched from a technical perspective. We could identify only a few articles dedicated to privacy impacts from a user's point of view. Nonetheless, together with our own results, the literature review sketched users out to be ambivalent towards connected vehicular services. While viewing the integration of the car into the smart grid as a promising opportunity for enhanced functions, users had severe privacy concerns. As Eyssartier (2015) reported, users do not want their private car to be equipped with smart sensors. During the course of this paper, we identified important factors that influence these concerns and thus the acceptance of connected vehicular services. Next to data type, the reviewed studies reported the purpose of data collection and the identity of the data-receiving party to be influential. Our own results allowed a closer look on some of these factors and detailed the acceptance of single data types as well as the trust in different data-receiving parties. However, we identified possibilities for manufactures and service providers on how to foster the adoption of connected vehicular services. Besides a perceived benefit from the usage of the service and monetary incentives, transparency in data collection and data processing motivates users to disclose their data.

User acceptance is indispensable for the successful introduction of new technologies (Petter et al. 2012). This especially applies for connected vehicular services that introduce pervasive computing into the private car (Schmidt et al. 2010) and thus might turn a so far private refuge transparent. Developers need to be particularly cautious when designing these systems or services. Highlighting important privacy-relevant aspects that should be taken into account, this paper provides first suggestions on how to configure those connected vehicular services in order to comply with user expectations.

First, collect your data parsimoniously. Though data have become a new source of high economical value (Tene and Polonetsky 2013), developers should not be parsimonious from a legal point of view. User views a broad range of data that are available within the vehicle as being sensitive. Collecting this data is likely to lower the acceptance of a new service.

Second, if this data is collected, share it with as few as possible. Our review showed that if more parties are granted access to the collected data, the lower user acceptance becomes.

Third, differentiate between private applications of the connected vehicular service in private versus professional contexts. Users tend to be more reluctant towards connected vehicular services in their private car than in professionally used cars. Therefore, privacy-relevant factors become even more relevant when designing a connected system or service for private cars.

Finally, communicate the extent and purpose of data collection transparently. User expects an honest and transparent communication of intention and purpose of data collection. As studies have identified transparency as an important success factor of information systems (Elia 2009), it should be a core interest of manufactures and service providers to design transparent connected vehicular services. Here, one might characterize transparency by the honest communication of privacy-relevant factors that have been found to be important for user acceptance: data type and identity of data receiver(s) & purpose of data. Moreover, during the course of this paper, we identified transparency as a strong incentive for users to disclose their data. Transparency builds trust between consumers and firms (Kang and Hustvedt 2014) and thus is a powerful tool to gain consumers.

References

Benedettini O et al (2009) State-of-the-art in integrated vehicle health management. Proc Inst Mech Eng Part G J Aerosp Eng 223(2):157–170

Chen L, Englund C (2016) Cooperative intersection management: a survey. IEEE Trans Intell Transp Syst 17(2):570–586

Chui KT et al (2016) An accurate ECG-based transportation safety drowsiness detection scheme. IEEE Trans Ind Inf 12(4):1438–1452

De Winter JC et al (2014) Effects of adaptive cruise control and highly automated driving on workload and situation awareness: a review of the empirical evidence. Trans Res Part F Traffic Psychol Behav 27:196–217

Dokic J et al (2015) European roadmap smart systems for automated driving. European Technology Platform on Smart Systems Integration, EPoSS. http://www.smart-systems-integration.org/public/news-events/news/eposs-roadmap-smart-systems-for-automated-driving-now-published

Dotzauer M et al (2015) Behavioral adaptation of young and older drivers to an intersection crossing advisory system. Accid Anal Prev 74:24–32

Derikx S et al (2016) Can privacy concerns for insurance of connected cars be compensated? Electr Mark 26(1):73–81

Dinev T, Hart P (2006) An extended privacy calculus model for e-commerce transactions. Inf Syst Res 17(1):61–80

Elia J (2009) Transparency rights, technology, and trust. Ethics Inf Technol 11(2):145–153

Endo T et al (2016) Study on privacy setting acceptance of drivers for data utilization on connected cars. In: 14th Annual conference on privacy, security and trust (PST), pp 82–87

Eriksson L, Bjørnskau T (2012) Acceptability of traffic safety measures with personal privacy implications. Trans Res Part F Traffic Psychol Behav 15(3):333–347

Eyssartier C (2015) Acceptability of driving an equipped vehicle with drive recorder: the impact of the context. IET Intel Trans Syst 9(7):710–715

Gardner B, Abraham C (2007) What drives car use? A grounded theory analysis of commuters' reasons for driving. Trans Res Part F Traffic Psychol Behav 10(3):187–200

Huang SC et al (2016) Smart Car. IEEE Comput Intell Mag 11(4):46–58

Jensen M et al (2016) Datenschutz im Fahrzeug der Zukunft: vernetzt, autonom, elektrisch, In: Lecture Notes in Informatics, p 441

Kang J, Hustvedt G (2014) Building trust between consumers and corporations: the role of consumer perceptions of transparency and social responsibility. J Bus Ethics 125(2):253–265
Li S et al (2011) Model predictive multi-objective vehicular adaptive cruise control. IEEE Trans Control Syst Technol 19(3):556–566
Müller A et al (2017) Einflussfaktoren auf die Akzeptanz des automatisierten Fahrens aus der Sicht von Fahrerinnen und Fahrern, In: 8. Darmstädter Kolloquium 7/8. März 2017 Technische Universität Darmstadt, 1
Papadimitratos P et al (2009) Vehicular communication systems: enabling technologies, applications, and future outlook on intelligent transportation. IEEE Commun Mag 47(11):84
Petter S et al (2012) The past, present, and future of IS success. J Assoc Inf Syst 13(5):341
Pohl J et al (2007) A driver-distraction-based lane-keeping assistance system. Proc Insti Mech Eng Part I J Syst Control Eng 221(4):541–552
Rebolledo-Mendez G et al (2014) Developing a body sensor network to detect emotions during driving. IEEE Trans Intell Transp Syst 15(4):1850–1854
Schmidt A et al (2010) Driving automotive user interface research. IEEE Pervas Comput 9(1):85–88
Schoettle B, Sivak M (2014) A survey of public opinion about connected vehicles in the US, the UK, and Australia. ICCVE 687–692:2014
Sheller M (2004) Automotive emotions feeling the car. Theory Culture Soc 21(4–5):221–242
Steg L (2009) Car use: lust and must Instrumental, symbolic and affective motives for car use. Trans Res Part A Policy and Pract 39(2):147–162
Tene O, Polonetsky J (2013) Big data for all: Privacy and user control in the age of analytics. Nw J Tech Intell Prop 11(5):239–273

MIX
Papier aus verantwortungsvollen Quellen
Paper from responsible sources
FSC® C105338

If you have any concerns about our products,
you can contact us on
ProductSafety@springernature.com

In case Publisher is established outside the EU,
the EU authorized representative is:
Springer Nature Customer Service Center GmbH
Europaplatz 3, 69115 Heidelberg, Germany

Printed by Libri Plureos GmbH
in Hamburg, Germany